ARK PAPE

THE AWAKENING EARTH

This impressive, stimulating tour de force is an exciting challenge to our traditional view of ourselves and our place in the universe. Peter Russell puts forward the idea of the earth as a collective self-regulating living organism, and invites us to join in an exploration of humanity's potential as seen through the eyes of the planet. Through this we share with the author an inner vision of our evolutionary future. Drawing on the work of physicists, psychologists, philosophers and mystics, Peter Russell shows that human society may be about to make a quantum leap in evolution as significant as the emergence of life itself some three and a half million years ago.

PETER RUSSELL

Peter Russell was an honorary scholar at Gonville and Caius College, Cambridge, where he studied mathematics, theoretical physics, experimental psychology and later computer science. He has undertaken research on the psychology of meditation and has appeared frequently on radio and television. Amongst his publications are *The TM Technique* and *The Brain Book*. In addition to teaching meditation, Peter Russell lectures frequently in Europe and the United States, and acts as a major consultant on the development of learning processes to several major international companies.

ARK

PETER RUSSELL
THE AWAKENING
EARTH
THE GLOBAL BRAIN

ARK PAPERBACKS

London, Melbourne and Henley

First published in 1982
ARK edition 1984
ARK PAPERBACKS is an imprint of
Routledge & Kegan Paul plc

39 Store Street, London WC1E 7DD, England

464 St Kilda Road, Melbourne,
Victoria 3004, Australia and

Broadway House, Newtown Road,
Henley-on-Thames, Oxon RG9 1EN, England

Printed in Great Britain
by Cox & Wyman, Reading

ISBN 0–7448–0012–9

Contents

Invitation

I would like you to come with me on a great adventure – an exploration of humanity's potential as seen through the eyes of the planet – and to share with me a vision of our evolutionary future. The journey will take us beyond this place and time, allowing us to stand back and behold humanity afresh, to consider new ways of seeing ourselves in relation to the whole evolutionary process.

We shall see that something miraculous may be taking place on this planet, on this little blue pearl of ours. Humanity could be on the threshold of an evolutionary leap, a leap which could occur in a flash of evolutionary time, and a leap such as occurs only once in a billion years. And the changes leading to this leap are taking place right before our eyes, or rather right behind them – within our own minds.

Put as bluntly as this, the hypothesis may seem an unbelievable fantasy. Yet I hope to show that it is a very real possibility. It is, moreover, a possibility which an increasing number of people are beginning to take seriously.

The seeds of my own explorations in this area were sown some twenty years ago while at school. I can recall lying in bed one night considering the rapidly increasing human population and the many ways in which we were consuming scarce resources and misusing the planet. It was no great effort to extrapolate these trends into the future and see that sooner or later impossible situations would occur. (To take an obvious example, there would eventually come a time when there would be more people alive than it was physically possible to feed.) Impossible situations do not occur. Therefore, I reasoned, before such points in time were reached, humanity would experience some very dramatic changes. Whatever happened we could not continue on the same path much longer.

In retrospect the conclusion is hardly profound, but for me it was an important turning point. It became very clear that sometime in my life I would probably witness the end of a set of trends which had been going on for thousands of years.

How would the changes come? At the time my attention was occupied with various 'negative' scenarios – nuclear holocaust, ecological collapse, worldwide famine, plagues or some as yet unforeseen catastrophe. These all seemed quite possible ways in which humanity's growing size and consumption could be curtailed, halted or even reversed.

But gradually over the years another far more optimistic scenario began to dawn. Rather than humanity suffering major set-backs, the dramatic change could be a growing-up and maturing of our species.

By this time I was at university studying theoretical physics. Fascinated as I was by science, however, I was even more fascinated by my own mind, and the minds of others. Western philosophy and psychology offered a few insights into the workings of the mind, but I had felt for a long time that there was a vast amount of wisdom locked up in the East, in particular in various teachings on meditation. So I ended up spending a winter in the Himalayan foothills studying with Maharishi Mahesh Yogi, and experiencing dimensions to my consciousness of which I had never dreamed. As a result I knew beyond any doubt that if everyone could contact such states of consciousness the world would be transformed. Humanity could change its direction constructively, rather than be changed destructively. So I returned to England and spent much of the next few years teaching meditation, encouraging others to discover for themselves a different way of being.

My vision of a transformed world continued to evolve, though I felt very much on my own, and often wondered whether it was all 'crazy'. Then one day a friend introduced me to the works of Teilhard de Chardin. Here was someone who had considered similar ideas about humanity's future, considered them far more deeply, and, more importantly, not been universally dismissed. I felt both inspired and strengthened.

From then on support started coming from many different directions; from developments in a number of sciences, from the writings of philosophers and visionaries both Eastern and Western, from conversations with others, and from my own experiences and insights. Piece by piece the jig-saw was coming together, and an overall picture began to emerge. More and more it appeared that we alive today may be standing on the threshold of an evolutionary development as significant as the emergence of life on Earth some 3,500 million years ago.

The nature of this possible transformation and the ways in which it could come about are what I want to explore with you in this book.

Our inquiry will draw upon the insights and experiences of many individuals, from mystics and religious teachers to scientists and astronauts and on recent developments in many different disciplines. Biology, chemistry, physics, astronomy, psychology, physiology, medicine, sociology and systems theory will all shed their relevant lights. (But do not fear, my aim is to share a vision, not to present an academic text, and where science comes in I shall keep it as simple as possible).

At times we will be looking at the similarities between aspects of society today and various phenomena in these sciences. In most cases these are not just analogies introduced to make a point clearer; they illustrate a deeper underlying pattern – what is technically called an *homology*. (The layout of bones in the forearm of a dog, elephant, seal and bat, for example, are in each case similar to the layout in the human forearm. This is an homology revealing a more fundamental common pattern.) When we start finding consistent underlying patterns running through the whole of evolution they give us a very strong reason for believing that society today may follow *homologous* developments.

No extrapolation into the future can ever be watertight, and the various examples and arguments are not meant to constitute an infallible prediction. Rather, they are supporting evidence; they provide a context within which such an evolutionary leap appears a possibility, and worth exploring further.

My goal is to convey a vision as a whole. It is the overall picture that is important rather than specific details. You may not like or agree with every point; indeed I would not expect you to accept everything – a garden may be enjoyed as a whole without you having to agree on the placement of every shrub and flower. Nor need the picture that emerges for you be the same as mine. You will probably start making connections with your own knowledge and experience. That's fine. I don't want you to believe me. I want to set you thinking – thinking about positive alternative futures.

The vision I shall be sharing is certainly a highly optimistic one – some might even say Utopian – and I make no apologies for this. As will become clear later, the image a society has of itself can play a crucial role in the shaping of its future. If we fill our minds with images of gloom and destruction, then that is likely to be the way we are headed. Conversely, more optimistic attitudes can actually help promote a better world. A positive vision is like the light at the end of a tunnel which, even though dimly glimpsed, encourages us to step on in that direction. It is the attractive force of such a future with which I shall be concerned, rather than the oppressive darkness which now seems to surround us.

The trip will be thought-provoking. It will challenge our traditional perspectives of ourselves and our place in the Universe. Yet you will probably find it resonating with some of your deeper gut feelings about humanity and the planet. Above all I hope it will excite you.

Acknowledgments

When I began this book I thought I had the ideas and structure all sewn up, and that six months, or a year at the most, would see it finished. That was three and a half years ago, and during that time hardly a day has passed when the book has not occupied me in one way or another. My reading was always bringing to light more information; conversations with friends were sparking off new ideas and insights; and in my quieter periods new syntheses suddenly appeared. Rather than being written, the book evolved – which is perhaps appropriate for a book whose main theme is evolution.

The new data, the new inspirations, and the new syntheses could go on forever. But now, six editions after the first, the time has come to put down my pen, and let you, the reader, have a turn at playing with some of the ideas, and sharing in some of my excitement.

With a book such as this it is impossible to give credit to everyone who has helped in one way or another over the course of its gestation, not to mention the many who influenced my thinking long before the book was ever conceived. In particular, I am indebted to Maharishi Mahesh Yogi and the knowledge he has shared over the years. He was a crucial trigger to my own thinking, and without his wisdom and my experiences of meditation I probably would never have started this venture.

A number of other writers, both past and present, have inevitably influenced my work. Those I probably owe the most to are Teilhard de Chardin, Sri Aurobindo, Walter Stace, Lancelot Law Whyte, Alan Watts and Olaf Stapledon. Since I have not set out to write a scientific thesis, I have not punctuated every fact or example with its academic heritage; instead I have included a 'Further Reading' section at the end in which all the relevant books are listed, together with a short description of each one. I trust they will each be as great a source of inspiration to the reader as they have been to me.

I must also credit that ordering principle within the Universe which manifests to us as synchronicity. Many times in my conversations and readings I find the same ideas and insights cropping up in different minds around the planet. It seems that when the time is ripe an idea will sweep through the collective unconscious appearing simultaneously in many guises. In these situations it is impossible to credit any one person as the 'originator': it is the cosmic creative intelligence, the pulse of evolution, which should be credited.

Special thanks must go to my American publisher Jeremy Tarcher both for his enthusiasm, and for far exceeding any writer's expectations of a publisher in constantly making sure everything was as clear as it could be. A more considerate and caring publisher I have yet to meet. Then I must also thank Stephanie Bernstein, who flew from L.A. to England on behalf of Jeremy to spend two weeks helping me put the finishing touches to 'the final draft' (braving an English summer and my own style of 'simple' living). She ended up staying seven weeks, and as a result another final version emerged. Painful though the process was at times, we saw it through; and it was worth it. Meanwhile my English publisher waited patiently.

Many others helped at various times. Mary Douglas, who one day got me to stop talking about the book and actually start writing, encouraged me throughout. Guy Dauncey, Michael Carey, Ned and Tinker Beatty, Ruth Bender, Mark Brown, Karen Brown, John St.John, Trevor Williams, Jane Henry and Norrie Huddle each gave me valuable criticism and feedback on the typescript in various stages of its evolution. Marion Warr, in addition to giving constant feedback, was a continual source of support, helping me through the writer's blues with kindness, understanding and patience. Thanks also to the numerous other friends who encouraged me to keep going – usually with cries of 'I want a copy, now!' And last, but not least, I am especially grateful to Pat Masters who was prepared to work till all hours when I suddenly unloaded a pile of work upon her, and who, having typed and re-typed the book many times, now knows parts of it off by heart.

And finally, two comments on word usage. I have used 'billion' to mean a US billion, i.e. a thousand million, since this is increasingly becoming its international usage; and larger units such as a trillion are similarly the US numbers.

As regards the 'he' dilemma, current alternatives such as 'e', 's/he', 'hesh', 'co', 'tey', 'de', 'he'er', 'thon', 'jhe', 'per' or 'wen', and even the continual use of 'he or she', tend to break the reader's flow. So, in order to keep the reading as smooth as possible, I have used 'he' throughout in its adrogynous sense of 'he or she' – except where the masculine sense is clearly intended.

PROLOGUE

CHAPTER 1

The blue pearl

Once a photograph of the Earth, taken from the outside, is available . . . a new idea as powerful as any in history will be let loose.

Fred Hoyle (in 1948)

What is home? To a man visiting a neighbour across the road, home is his house, his rose bushes, his backyard. To the peasant bringing goods to the city, it is his village. To the traveller abroad, it is his country. These are common human experiences. At one time or another, we each have called our town, state, or nation: home. Yet there is a greater home which has only recently come into our awareness, even though it has been ours all along – planet Earth.

As the first astronauts travelled out into space, and the Earth receded into the distance, national boundaries began to lose their significance. These space pioneers found themselves no longer identified with a particular country, class or race, but with humanity and the planet as a whole. Standing on the lunar surface, the astronauts saw what no human being had ever seen before: the great sphere that is Earth, four times as big and five times as bright as the moon itself.

To Edgar Mitchell, the sixth man to stand on the moon, this was a deeply moving experience, and he felt a strong mystical connection to the planet.

It was a beautiful, harmonious, peaceful-looking planet,
blue with white clouds, and one that gave you a deep
sense . . . of home, of being, of identity. It is what I prefer to
call instant global consciousness.

Mitchell observed that everyone who has been to the moon
has had similar experiences. 'Each man comes back with a
feeling that he is no longer an American citizen – he is a
planetary citizen.'

Russell Schweickart, another astronaut, similarly felt a
profound change in his relationship to the planet.

You realise that on that small spot, that little blue and
white thing, is everything that means anything to you – all
of history and music and poetry and art and death and birth
and love, tears, joy, games, all of it on that little spot out
there . . . You recognise that you are a piece of this total
life . . . And when you come back there is a difference in
that world now. There is a difference in that relationship
between you and that planet and you and all those other
forms of life on that planet, because you've had that kind of
experience.

But the astronauts were not the only ones to have
experienced this profound change of perspective. The photo-
graphs of our planet brought back from space triggered
similar deep reactions in many Earth-bound men and women
– feelings of awe and connectedness. This is our home, seen at
last as a whole, in all its beauty and magnificence.

The profound impact of this Earthview has resulted in this
picture being used in almost every sphere of human activity.
It adorns the walls of offices and living rooms; it is a greeting
card, a T-shirt, a book cover and a bumper sticker. Ecological
movements and planetary organisations incorporate it in their
logos, as do educational institutions and business corporations. At one time or another it has been used to advertise just
about everything from cars, washing machines and shoes, to
book clubs, banks and insurance companies. Yet in spite of all
this exposure, the picture still strikes a very deep chord, and
none of its magnificence has been lost.

It is not entirely coincidental that this photograph achieved its widespread appeal at the same time as many people were becoming increasingly concerned about the relationship between humanity and the planet, and the need for us to live in harmony with each other and with our environment. The picture has become a spiritual symbol for our times. It stands for the growing awareness that we and the planet are all part of a single system, that we can no longer divorce ourselves from the whole.

So the most valuable spin-off from the moon expeditions may not have been in the fields of science, economics, politics or the military, but in the field of consciousness. Getting to the moon has allowed humanity for the first time in its history to look upon this blue pearl that has been our home for millions of years, and to see it as a whole. As Edgar Mitchell pointed out, 'the payoff from Apollo may be inestimably richer than anyone anticipated'.

THE LIVING EARTH

The view of Earth from space brought with it yet another insight: the realization that the planet as a whole may be a living being. We Earthlings might be likened to fleas who have spent all their lives on an elephant, unaware of what it really was. They charted the terrain – all the different patches of skin, hairs and bumps – studied the chemistry, plotted the temperature changes, and classified the other animals that shared their world, arriving at what they thought was a reasonable understanding of where they lived.

Then one day a few of the fleas took a huge leap and looked at the elephant from a distance of a hundred feet. Suddenly it dawned: 'The whole thing is alive!' This was the truly awesome realisation the trip to the moon brought to many. The whole planet appeared to be alive – not just teeming with life, but an organism in its own right.

If the idea of the Earth as a living being is initially difficult to accept, it may be partly due to our notions of what sort of things can and cannot be organisms. We accept a vast range óf systems as living organisms, from bacteria to blue whales, but when it comes to the whole planet we might balk a little. It is, however, worth reminding ourselves that four hundred years ago no one realised that there were organisms within us and around us, so small that they could not be seen with the naked eye. Only with the development of the microscope did people begin to surmise that there were living organisms that minute. Today we are viewing life from the other direction, through the 'macroscope' of the Earthview; and we are beginning to surmise that something as vast as our planet might also be a living organism.

This hypothesis is all the more difficult to accept because the living Earth is not an organism we can normally observe outside ourselves: it is an organism of which we are an intimate part. Only when we step into space can we begin to see it as a separate being. Stuck like fleas on an elephant, we have not, until recently, had the chance to see the planet as a whole. Would a cell in our own bodies, seeing a tiny part of the inside of the body for a short period, ever guess that the body as a whole was a living being in its own right?

Another reason we may find this a rather strange idea is that our everyday perspectives and thinking about the planet have generally been in timescales appropriate to human life. The planet's own timescale, however, is vastly greater than ours. The rhythm of day and night might be considered the pulse of the planet, one full cycle for every hundred thousand human heartbeats. Speeding up time appropriately, we would see the atmosphere and ocean currents swirling round the planet, circulating nutrients and carrying away waste products, much as the blood circulates nutrients and carries away waste in our own bodies.

Speeding it up a hundred million more times, we would see the vast continents sliding around, bumping into each other, pushing up great mountain chains where they met. Fine threadlike rivers would swing first one way then another, developing huge meandering loops as they accommodated themselves to the changes in the land. Giant forests and grasslands would move across the continents, sometimes thrusting limbs into new fertile lands and at other times withdrawing as climate and soil changed.

If we could look inside, we would see an enormous churning current of liquid rock flowing back and forth between the centre of the planet and the thin crust, sometimes oozing through volcanic pores to supply the minerals essential for life.

Had we senses able to detect charged particles, we would see the planet bathing not only in the light and heat of the sun, but also in a solar wind of ions streaming from the sun. This wind, flowing round the Earth, would be shaped by her magnetic field into a huge pulsating aura streaming off into space behind her for millions of miles. Changes in the Earth's fluctuating magnetic state would be 'visible' as ripples and 'colours' in this vast comet-like aura, and the Earth herself would be but a small blue-green sphere at the head of this vast energy field.

Thus if we look at the planet in terms of its own processes, we begin to see a level of complex activity reminiscent of that found in a living system. Such similarities do not, however, constitute any form of proof. The question we have to ask is whether scientists could accept the planet as a single organism in the same way they accept bacteria and whales? Could the Earth actually *be* a living organism?

This no longer seems so far-fetched. On the contrary, an increasingly popular scientific hypothesis suggests that the most satisfactory way of understanding the planet's chemistry, ecology and biology, is to view the planet as a single living system.

THE GAIA HYPOTHESIS

One of the major proponents of the theory that the planet behaves like a living system is British chemist and inventor Dr. James Lovelock. Interestingly, his ideas, which have fundamentally altered many people's perception of the planet, are another fortuitous spin-off from the space race.

In the early 1960s Lovelock served as consultant to a team at the California Institute of Technology working on plans for the investigation of life on Mars. One of the problems they faced in looking for alien lifeforms was that they did not know exactly what they were looking for. Other lifeforms might be based on completely different chemistries – on silicon rather than carbon, for instance – and might not reveal themselves to tests based on Earth-type life.

Lovelock theorized that however strange the chemistry and lifeform might be, there would be one very general characteristic: any lifeform would take in, process and cast out matter and energy, and this would have detectable effects upon the physical surroundings. Thus if a planet were devoid of life, the chemical constituents of the atmosphere, oceans and soil would, through their interactions over millions of years, settle into a state of equilibrium, and the proportions of the various constituents in this state could be roughly predicted by the laws of physical chemistry. If, on the other hand, some form of life were present, then, whatever chemical processes it was based on, it would almost certainly leave the environment in a state recognizably different from that predicted by physical chemistry alone.

As a very simple example of this principle we might consider a jar containing a mixture of sugar and water. Physical chemistry predicts that the sugar will dissolve until a given concentration is reached. If, however, life in the form of yeast cells were added and left to grow, the resultant mix would be very different: there would be a lower concentration of sugar than expected, and much higher levels of alcohol and other organic products. So we could get a good idea of whether there was (or had been) life in the jar by measuring the sugar and alcohol concentrations.

The beauty of Lovelock's approach to life detection is that one does not need to visit another planet to know whether or not life is there. The basic chemistry of the atmosphere can be deduced from Earthbound examination of the various infrared, light and radio waves coming from the planet. In the 1960s enough was known about the Martian atmosphere to suggest that it was very close to the chemical equilibrium state; it showed no signs of the exotic chemistry characteristic of the presence of life. So, Lovelock concluded, it was extremely unlikely that there was any life on Mars.

Applying a similar approach to the atmosphere, ocean and soil of our own planet, Lovelock found that the chemical constituents were far removed from the equilibrium predicted by physical chemistry. To the casual onlooker it might seem that he had merely shown that there was, after all, life on Earth. But Lovelock began to see far greater significance in these disequilibria.

Firstly, the concentration of gases in the Earth's atmosphere differ by factors of millions from the levels predicted by physical chemistry. For example, the predicted level of oxygen in the air would be virtually zero, yet the actual concentration is about 21%. This is puzzling because oxygen is a highly reactive gas, combining readily with many other chemical elements, and should therefore be rapidly absorbed. Secondly, and even more puzzling, the actual composition of the atmosphere has remained at a level that is the optimum for the continuance of life.

After considerable pondering upon many such highly unlikely characteristics, Lovelock came to 'the only feasible explanation': the atmosphere is being manipulated on a day-to-day basis by the many living processes on Earth. The entire range of living matter on Earth, from viruses to whales, from algae to oaks, plus the air, the oceans and the land surface all appear to be part of a giant system able to control the temperature and the composition of the air, sea and soil so as to maintain the optimum conditions for the survival of life on the planet. This concept Lovelock termed the *Gaia Hypothesis*, in honour of the ancient Greek 'Earth Mother' goddess, *Gaia* (or *Ge*). In this context Gaia signifies the entire biosystem – all the plants, animals and fungi living on the planet – plus the atmosphere, the oceans and the soil.

In maintaining the optimum conditions for life, Gaia manifests a characteristic which all living systems have in common: homeostasis. Coming from the Greek 'to keep the same', this term was first coined by Claude Bernard, a 19th-century French physiologist, who stated that 'all the vital mechanisms, varied as they are, have only one object: that of preserving constant the conditions of life'.

One example of homeostasis is the human body's maintenance of a temperature of about 37 degrees centigrade. This is the optimum temperature for the majority of the body's metabolic processes. Although the external temperature may vary by many tens of degrees, our internal temperature seldom varies by more than a degree or two, the body cooling itself through sweating and warming itself through physical activity and shivering. Other examples of homeostasis are the regulation of the number of white blood cells; the control of the acidity, salt content and delicate chemical balance of the blood; and the maintenance of a steady water balance by the kidneys. These and many other homeostatic processes together maintain the optimum internal environment for the continuance of our body's life processes. Such processes are not only found in the human body and in all living systems, but also within Gaia herself.

Gaia appears to maintain planetary homeostasis in a variety of ways, monitoring and modifying many key components in the atmosphere, oceans and soil. The data that Lovelock amasses in support of this contention is quite fascinating, but since it would take many pages to give it full justice I will not go into all the details (the interested reader should take a look at Lovelock's book, *Gaia. A New Look at Life on Earth*). Here, in summary, are some of the indications of Gaia's homeostatic mechanisms:

- The steadiness of the Earth's surface temperature: Although life is found to exist between the extremes of -5 and 105 degrees centigrade, the optimum range is between 15 and 35 degrees centigrade. The average temperature of most of the Earth's surface appears to have stayed within this range for hundreds of millions of years, in spite of drastic changes in atmospheric composition and a large increase in the heat received from the sun. (If at any time in the Earth's history the overall temperature had gone outside these limits life, as we know it, would have been extinguished.) Such behaviour is reminiscent of our bodies' maintenance of an optimal internal temperature despite large variations in the external temperature.

- The regulation of the amount of salt in the oceans: At present the oceans contain about 3.4 percent salt, and geological evidence shows that this figure has remained pretty constant, despite the fact that salt is being continually washed in by the rivers. If the salt concentration had ever risen as high as 4 percent, life in the sea would have evolved through very different organisms from those found in the fossil records. If it had gone beyond 6 percent, even for a few minutes, life in the oceans would have immediately come to an end, for at this level of salinity cell walls disintegrate – cells would literally fall apart. The oceans would have become like the Dead Sea – an intolerable environment for life.

- The stabilization of the oxygen concentration of the atmosphere at 21 percent: This is the optimum balance for the maintenance of life: a few per cent less and the larger animals and flying insects could not have found enough energy to survive; a few per cent more and even damp vegetation would burn well. (A forest fire started by lightning would burn fiercely and indefinitely, eventually burning all vegetation on the Earth's land surface.)
- The presence of a small quantity of ammonia in the atmosphere: This is just the amount needed to neutralise strong sulphuric and nitric acids produced by the natural combination of sulphur and nitrogen compounds with oxygen (thunderstorms, for example, produce tons of nitric acid). The net result is that the rain and soil are kept at the level of acidity optimum for life.
- The existence of the ozone layer in the upper atmosphere: This shields life on the surface from ultra-violet radiation, which damages the molecules essential for life, particularly the DNA molecules found in every living cell. Without it life would very rapidly be annihilated.

On the basis of these and other 'homeostatic' behaviours, Lovelock concludes that the climate and chemical properties of the Earth seem always to have been optimal for life as we know it.

Critics of the Gaia Hypothesis might argue that the origin and maintenance of life on this planet resulted from a series of very lucky coincidences. If, for example, the proportion of ammonia in the early atmosphere had been a little higher or lower, the Earth would have ended up too hot or too cold for life to begin. They might argue that it has been a series of flukes that kept the planet's surface temperature roughly constant while the sun's output changed; a series of flukes that has kept the levels of carbon dioxide, oxygen, salt and many other chemicals at the optimum levels for the maintenance of life; and a fluke that there is an ozone layer to protect us from lethal quantities of ultra-violet light.

In the same way, a cell in the human body, observing the body's continued survival through hot, cold, and many other changes, might, if it were so inclined, put it all down to a series of lucky coincidences: the body just happens to sweat when it is hot, just happens to shiver when it is cold, just happens to take in the right amount of nutrients when they are needed. Perhaps it is a fluke that the blood sugar, acidity and salinity stay at the optimum levels, and that red blood cells happen to bring along oxygen and take away the waste. From such a point of view, the body survives from one moment to the next as a result of an extremely fortunate series of coincidences.

This clearly is not the case. The body behaves in a well-ordered manner with a definite sense of purpose. It sweats, shivers, eats, breathes and regulates its internal functions and chemical constituents in order to preserve a state of homeostasis, and so survive.

Just as this makes more sense of the body's activities, so it makes more sense of the planet's. Gaia appears to be a self-regulating, self-sustaining system, continually adjusting its chemical, physical and biological processes in order to maintain the optimum conditions for life and its continued evolution.

Can we therefore consider the biosphere as a single living organism? Lovelock is cautious on this point; he sees the atmosphere to be like a beehive or a cat's fur – a biological construction designed to maintain a chosen environment, though not actually living in itself. This may be true of the atmosphere considered in isolation, but is it equally true of the entire biosphere, in which the atmosphere is an integral part? A cat's fur may not in itself be living, but it is nevertheless part of the cat. Without its fur a cat would be a different creature, with different bodily processes. If we take the atmosphere, oceans and soil to be an intrinsic part of the complete biosystem, can we consider the system as a whole to be alive?

Before we can answer this question we need to look more closely at the general characteristics common to all living systems and see to what extent Gaia satisfies them.

GENERAL LIVING SYSTEMS THEORY

Until the middle of this century, each scientific subject was treated more or less as an isolated area: physiologists studied the body, sociologists studied social groups, and engineers studied mechanical systems. Each discipline had its own theories and understandings, and these generally had very little connection with the findings of other sciences.

In the late 1940s, biologists such as Ludwig von Bertalanffy and Paul Weiss began to remedy this by focussing on the common principles and properties underlying widely different phenomena. The concept of homeostasis, for example, originally applied to physiological processes, was extended by von Bertalanffy to encompass a much wider range of phenomena – from single cells to whole populations. Likewise, the concept of feedback, which originally came from engineering, was seen to be applicable to physiological, psychological and social phenomena. The insights gained from developing general models provided the impetus for the development of the interdisciplinary study now known as General Systems Theory.

The term 'theory' is in fact rather misleading. General Systems Theory is not so much a specific theory as a way of looking at the world. It sees the world as an interconnnected hierarchy of matter and energy. According to this view, nothing can be understood on its own; everything is part of a system (a system being defined in its most general sense as a set of units which are related to each other and interact). Systems may be abstract, as in mathematical systems and metaphysical systems, or concrete, as in a telephone system or transport system.

One branch of General Systems Theory deals particularly with living systems. In his magnum opus, *Living Systems*, James Miller, one of the pioneers of this approach, proposed that all living systems are composed of subsystems which take

Table 1. The nineteen subsystems of a general living system, with examples at the level of the human being, a country and the biosphere. (The examples given are not exhaustive; many others may be found.)

SUBSYSTEM	LEVEL Human Being	Human Society (a nation)	Biosphere (Gaia)
Ingestor: Brings matter-energy across boundary from outside	Mouth, nose and lungs	Import company, Airlines	Atmosphere (transparent to visible light and infra-red, and permitting cosmic dust to fall through) Volcanoes (permitting flow of minerals through Earth's crust)
Distributor: Carries matter-energy around the system	Blood	Transport company, Oil pipeline	Temperature and pressure gradients in atmosphere and oceans, Animal migrations and wandering
Converter: Changes certain inputs into more useful forms	Teeth, stomach, small intestine, liver, pancreas	Oil refinery, Farm	Mosses and lichens converting minerals to humus, Plants photosynthesising light into chemical bonds
Producer: Forms stable associations among inputs, or outputs of converter, for growth, repair, movement	Protein synthesis by RNA, Production of new skin by epidermis	Factory, Construction company	Producers occur at cellular level, e.g. chloroplasts, mitochondria, RNA, and in reproduction of each species
Matter-Energy Storage:	Fatty tissues, Calcium in bones	Warehouses, Dammed rivers	Dead plant and animal matter in soil, Water in oceans and atmosphere
Extruder: Transmits waste matter-energy out of system	Urethra, Anus, Lungs	Export company, Smokestacks, Refuse collectors	Sedimentation in oceans, Gaseous escape through upper atmosphere
Motor: Moves system, or parts of it, or moves environments	Muscles	Cars, trains, boats, planes	Tides, Climate changes, Continental drift

Table 1 continued

Supporter: Maintains proper spatial structure	Skeleton	Housing, Public buildings	Earth's crust, Buoyancy of air and sea
Input Transducer: Sensory receptors for information coming from outside	Eyes, Ears, Heat sensors	Foreign news service, Scientific research	Animals and plants reacting to day and night, to seasons and earthquakes
Internal Transducer: Receives information about changes going on within system	Hypothalamus in brain monitoring temperature, salt content of blood	Public opinion polls, Political parties	Animal and plant reactions to changing climate, floods, aridity, pollution
Channel and Net: Routes by which information transmitted to all parts of system	Central and peripheral nervous system, Hormonal system	Books, magazines, telephones, Postal services, conferences	Animal migration and wandering, Seed dispersal in plants, Availability of food
Decoder: Translation of input information into internal meaningful code	Retina of eye, Visual cortex of brain	Translators, Commentators, Foreign Office	Interspecies communication – response to reactions of other living beings
Associator: Associates items of information, the first stage of the learning process	Temporal and frontal lobes of the brain	Scholars	Changed habitats and behaviours
Memory: Stores various types of information over different periods of time	Entire brain	Libraries, Data bank	Evolutionary adaptations recorded in changed genes

(continued)

15

Table 1 continued

Decider: Receives information from other subsystems and transmits to them information controlling entire system	Various brain centres, Spinal chord, Pituitary gland	Governments, Law courts, Voting public	Soil, Interspecies communication
Encoder: Translates internal information to external messages	Speech area of brain	Newspapers	Changes in constituents of atmosphere
Output Transducer: Changes information into other matter-energy forms and transmits them into environment	Voice box, facial expressions	TV station, Official spokesman	Upper atmosphere, gaseous loss and radiation, Changed albedo (reflectivity) of planet
Reproducer: Gives rise to other similar systems	Sexual organs	Settlers abroad, Social reformers	(The biosphere has not (yet) displayed this characteristic), Viruses lost to space? Interplanetary travel?
Boundary: Holds system together, protects from external stresses, excludes or permits various inputs and outputs	Skin	Customs officials, National border	Earth's crust below, Upper atmosphere above

in, process, and put out matter, energy, or information, or combinations of these. He identified nineteen critical subsystems that seem to characterize living systems.

The first eight subsystems are concerned with matter-energy processes and essentially portray the way any living system ingests, digests, uses and excretes physical matter and energy. All living systems, for example, have an *ingestor* – some means of taking in matter and energy – whether it be a gap in a cell wall, an artery leading into an organ, the mouth of an organism, or a major seaport. The next nine subsystems are concerned with information processes – the ways in which living systems sense the environment, abstract information, integrate it and remember it. One such subsystem is the *input transducer*, which brings information into the system. This may be the receptor site in the membrane of a nerve cell, the eye of an organism, or the foreign news service of a nation. The last two subsystems, the *reproducer* and the *boundary*, are processes involving both matter-energy and information processes. The reproducer leads to the creation of new systems similar to its own, through the transmission of both physical matter and information about the original system. The boundary holds the whole system together, excluding or permitting the entry or exit of various types of matter, energy and information.

Looking at the entire biosystem from the perspective of General Living Systems Theory, we find each of the nineteen critical subsystems at work. The ingestors, for example, are the upper atmosphere, through which the sun's energy and cosmic dust are taken in, and the Earth's crust, through which minerals well up. The input transducers are the many plants and animals as they react to daily and seasonal changes, or to earthquakes and sunspot activity. Table 1 outlines all the nineteen subsystems as they apply to the human body, and to the biosystem of planet Earth. (It also shows how they apply to a human society – a point we shall be returning to later.)

But does the finding that the biosystem appears to possess each of the nineteen subsystems characteristic of life prove that it is indeed a living system? Miller makes a strong case to show that these subsystems are all necessary characteristics (even though some of the subsystems are not always easy to identify – we are still not clear how memories are stored either in the cell or in the human brain), but are they sufficient? The answer is almost certainly 'No'. An automobile displays many of these characteristics, and with various modifications and additions could be made to satisfy them all, including reproduction if we so desired, but it would very clearly not then become a living system.

There is one more characteristic common to all living systems, which clearly distinguishes them from non-living systems. This is a living system's ability to maintain a high degree of internal order despite a continually changing environment – something we shall be looking at in depth in Chapter 3. Our bodies maintain the same basic structure in a variety of conditions and will strive to repair themselves when damaged. We adapt to changes and learn from experience. Machines, however, do not generally show this characteristic. They wear out and run down; they are not self-organising.

It is very difficult to find examples of non-living systems which both possess the nineteen critical subsystems *and* are self-organising. At present, therefore, satisfaction of both these criteria seems to be a reasonable sufficient condition for a system to be a living system.

Gaia appears to satisfy both criteria. Its self-organising nature has already been clearly demonstrated in Lovelock's work on the biosystem's ability to maintain planetary homeostasis. It also satisfies Miller's criteria. Taken together, these two findings argue strongly that Gaia should rightly be considered as a living system in its own right.

HUMANITY IN GAIA

If the entire biosphere has evolved as a single living system, in which all the numerous subsystems play diverse and mutually dependent roles, then humanity, being a sub-system of this larger planetary system, cannot be separated from it or treated in isolation. What then is its function in relationship to Gaia?

There seem to be two common and opposing responses to this question. The first is that humanity is like some vast nervous system – a global brain in which each of us are the individual nerve cells. The second, more pessimistic, possibility is that we are like some kind of planetary cancer.

Considering the first response, human society, like our own brain, can be seen as one enormous data collection, communication and memory system. We have grouped ourselves into clusters of cities and towns rather like the way nerve cells cluster into ganglia in a vast nervous system. Linking the 'ganglia' and the individual 'nerve cells' are vast information networks.

Society's slow package systems of communication such as the postal services, with specific items sent to different parts of the system, are like the relatively slow chemical communication networks of the body, such as the hormonal system. Our faster, electronically based telecommunication networks (telephones, radio, computer links, etc.) are like the billions of tiny fibres linking the nerve cells in the brain.

At any instant there are millions of messages flashing through the global network, just as in the human brain countless messages are continually flashing back and forth. Our various libraries of books, tapes and other records can be seen as part of the collective memory of Gaia. Through language and science we have been able to understand much of what happens around us, monitoring the planet's behaviour much as the brain monitors the body's. We might see the Western and Eastern cultures as the two sides of Gaia's brain – the rational/intellectual left and the more intuitive right. And humanity's search for knowledge could be Gaia's way of knowing more about herself and about the universe in which she lives.

Many of the above parallels relate to the higher mental functions, to thinking, knowing, perceiving and consciousness, to the functions associated with the cortex of the human brain – the thin layer of nerve cells wrapped around the outside of the brain – and it might be more accurate to liken humanity to the cortex of the planet.

In evolutionary terms, the cortex is a relatively late addition, most of its development occurring with the mammals. It is not necessary for the maintenance of life; the cortex of an animal can be removed, yet the heart, lungs, digestion and metabolism will continue. In a similar way the planet

Earth has survived perfectly well without humanity for over 4,000 million years, and could continue very well without it.

This brings us to the second possibility – that humanity might be some form of recently-erupted malignant growth, which the planet would be better off without. This possibility occurred to Edgar Mitchell while standing on the moon. Immediately after feeling a sense of identity with the planet as a whole, came the opposite feeling, 'that beneath that blue and white atmosphere was a growing chaos that the inhabitants of planet Earth were breeding among themselves – the population and technology were growing rapidly out of control. The crew of 'spacecraft Earth' was in virtual mutiny to the order of the Universe.'

The analogy with cancer cannot be ignored. Modern civilisation seems to be eating its way indiscriminately across the surface of the planet, consuming in decades mineral resources which Gaia herself inherited billions of years ago. At the same time humanity is threatening to destroy the biological fabric which took millennia to create. Large forests essential to the ecosystem are looking moth eaten, animal species are being hunted out of existence, lakes and rivers are turning sour, and large areas of the planet are being laid barren by mining and the spread of concrete. Indeed, an aerial photograph of almost any large metropolis with its sprawling suburbs is very reminiscent of the way some cancers grow in the human body. Technological civilisation really does look like a rampant malignant growth blindly devouring its own ancestral host in a selfish act of consumption.

This view might seem to be in opposition to the idea of humanity being some form of global brain. It is entirely possible, however, that both these perspectives of humanity's role in Gaia may be valid. Perhaps we are part of some global nervous system, currently passing through a very rapid phase of development, capable of being to the planet everything that our own brains are to us. Yet this nervous system has, at a very critical stage, appeared to have gone out of control, threatening to destroy the very body which supports its existence.

If then we are to fulfil our role as a part of the planetary brain, our malignant behaviour must be stemmed and the negative trends reversed. If we are to achieve this, it is imperative that we change, in the most radical way, our attitudes towards ourselves, others and the planet as a whole. As we shall be seeing, such changes are going to require a major transformation in human consciousness. To appreciate what such a transformation could mean, both for humanity and for Gaia, and how the changes could come about, it will be valuable to look first at our past, at the whole evolutionary process. As we look at the basic principles which underlie it, and where it has come to, we may gain a clearer idea of where it is headed.

PART ONE

Evolution Past, Present and Future

CHAPTER 2

Evolution so far

Matter has reached the point of beginning to know itself . . .
(Man is) a star's way of knowing about stars.

George Wald

What do we mean by evolution? To most people the word
probably signifies the gradual development of the many living
species, one from another, probably along the lines that
Charles Darwin proposed in his treatise *On the Origin of Species*.
Here, however, we shall be considering evolution in a far
broader context than the origin and development of life. We
shall be going back in time to consider the unfolding of the
Universe long before life appeared, and looking at the origin
and development of matter itself, for without this earlier
evolution life could never have started. And we shall be
stepping into the future to explore what developments might
lie beyond the evolution of human beings. From this
perspective the evolution of life is but an act in a far grander
cosmic play. But let us start at the beginning, with the origins
of the Universe itself.

How did the Universe begin? Did it even have a beginning?
Many theories have been put forward at different times, some
based on physical science, others based on spiritual, metaphysical or philosophical frameworks. At present the most
widely accepted theory in the West is the scientific model
based on the notion that the Universe started with a 'Big

Bang' some fifteen billion or so years ago. According to this view, the whole Universe as we know it, was born from a gigantic superhot fireball, which rapidly expanded and cooled, condensing over billions of years into countless galaxies and myriads of stars.

Of what happened before the Big Bang, physical science knows nothing, and may forever know nothing. Time and space only came into being once the process began – hard as that may be for us to grasp. Science also knows virtually nothing about what happened in the first one hundredth of a second of the Big Bang. The temperature of the Universe was well over 1,000,000,000,000 degrees centigrade, so hot that electrons, protons and other elementary particles could not exist. There is no known physics which can describe what happens in such a superhot state. The most we can say about the Universe during this time is that it was a state of pure energy, a Universe dense with electromagnetic radiation. To borrow from another view of creation – in the beginning, there was Light.

Current theories state that after the first one hundredth of a second, the Universe had cooled down sufficiently (to 100,000,000,000 degrees centigrade) for elementary particles – electrons, protons and neutrons – to form. At this stage, the Universe was expanding very fast, and cooling rapidly as it did so. Yet it was still far too hot for any simple atomic nuclei to form, let alone complete atoms. Any particles which did by chance come together to form nuclei would have been instantly shaken apart by the intense heat energy. At this stage nuclei were annihilated as fast as they were created. Only after about three minutes had elapsed, when the Universe had cooled down to about 900,000,000 degrees centigrade, could neutrons and protons combine to form stable atomic nuclei – initially the nuclei of hydrogen and helium.

The Universe continued to expand and cool until, after about 700,000 years, the temperature had decreased to around 4,000 degrees centigrade, which is about the temperature of our sun. At this temperature, electrons and nuclei could stay together and so form complete simple atoms.

At the same time as the intense radiant heat energy was

driving matter apart, the much weaker gravitational attraction of matter for matter was tending to pull it back together. Below 4,000 degrees centigrade, the pressure of the radiant heat had decreased to the point where gravitational effects began to dominate and atoms began to clump together. Wherever the atoms happened to be a little more densely clustered, they produced a slightly stronger gravitational field, attracting other atoms towards them. Slowly the irregularities were amplified and, over thousands of millions of years, these eddies became clusters of primordial galaxies. Within these giant clouds the hydrogen and helium gases continued to gather in ever more condensed masses, eventually giving birth to the first stars.

By this time the Universe as a whole had become much colder, even by human standards – a few tens of degrees above absolute zero. As the cool gas was drawn into stars, however, it gained energy from gravitational collapse. This, combined with the energy of the star's own radioactivity, heated the gas up again to several million degrees.

Many of these early stars generated so much heat that they eventually flared up and exploded in brilliant supernovae, each as bright as an entire galaxy. The huge quantities of energy generated made it possible for nearly all the other chemical elements to form. Current theories of stellar evolution suggest that this process would have occurred as part of a thermonuclear chain reaction, and that within such supernovae 15 percent of the heavier elements would have been formed in about 10 seconds.

The force of the explosion sent these heavier elements spewing out into space. Slowly, over millions and millions of years, the debris condensed into new stars, forming yet more complex atoms. Later, these stars also exploded, spewing their material out into space. This process has been repeated several times over the past 15 billion years, and it is thought that our own sun is probably a fourth generation star.

One consequence of this recycling and regeneration of matter is that every atom on this planet (with the possible exception of some hydrogen and helium left over from the Big Bang) has been processed in at least one star. Virtually every atom in your body has at some stage in its long history passed through or been created in one of these giant stellar furnaces. As a corollary, the chemical composition of our planet was fixed at its birth (with the exception of some minor changes though radioactive decay, and some cosmic dust falling in from space). The atoms which today constitute your body may in their past have been in a volcano, in rocks, in the oceans, in the atmosphere, in an oak tree, in an eagle, and in other people both past and present. What has changed over the eons are the combinations the atoms have made with each other, not the atoms themselves.

THE BUILDING BLOCKS OF LIFE

Before life could emerge, atoms had to combine to form molecules of increasing complexity. The stars themselves were far too hot for even the simplest molecules to form; the energy was such that any atoms that did come together were immediately flung apart again. The formation of a large variety of stable molecules needed the more moderate temperature ranges found in the regions around the stars. Here, on the cooler planets, and perhaps in the surrounding space itself, atoms could begin to combine with each other to form simple molecules, substances such as water, carbon dioxide and salts of various kinds. Our own solar system was probably formed about 4.6 billion years ago, from a huge cloud of interstellar dust. Most of the cloud was composed of frozen hydrogen, helium and ice, but the planet Earth was fortunate to condense out of a part of the cloud rich in a diversity of elements, including all those necessary for the evolution of carbon-based life.

How life actually began is still a matter of considerable debate. The most popular model supposes that the early atmosphere consisted of a mixture of hydrogen, ammonia, methane, carbon dioxide, hydrogen sulphide, water vapour and other simple gases formed from combinations of the lighter atoms. These gases, it is hypothesised, could have combined to form the various chemical compounds essential for life. These combinations are stable at temperatures below the boiling point of water, and they could have formed as soon as the Earth's surface temperature fell below this level, some 4 billion years ago – which, on the Earth's timescale was not long after its birth.

In a now famous experiment, performed in 1953 by Stanley Miller, then a graduate student at the University of Chicago, a 'primeval soup' of water, methane, nitrogen, ammonia and traces of hydrogen was subjected to electric sparks (simulating lightning). Within hours a large variety of organic substances such as sugars, aldehydes, carboxylic acids and amino acids were formed, chemical compounds which in fact are some of the basic constituents of all forms of life known on this planet.

The experiment is so simple that it has since been repeated hundreds of times, even by high school students, with similar results. By varying the proportions of the different gases present, and by substituting ultraviolet light for electric discharges, subsequent researchers have found that it is possible for all the basic building blocks of life to be created by such processes. Moreover, they can be created under a variety of different conditions. Further experiments have shown that it is not even necessary to have a methane and ammonia-rich atmosphere. The same molecules can also be built up in atmospheres rich in carbon dioxide, and even in the icy cold of frozen oceans.

The fact that these chemicals are so easily created, and under so many different conditions, suggests not only that it was almost bound to happen, but also that it could have happened almost anywhere. Wherever these conditions were met – and there probably have been billions of planets in the Universe which passed through similar stages – the basic building blocks of life were almost certainly created.

Furthermore, their formation is not necessarily limited to planets. Experiments have shown that these basic molecules can even form in a near vacuum, at temperatures approaching absolute zero (the conditions found in interstellar space) and recently many of these compounds have indeed been detected far out in space.

That hydrogen and even helium occur throughout interstellar space has been known since the early days of radio astronomy, but until recently few people thought that more complex molecules existed out there. In 1965, however, cyanide (one atom of carbon connected to one of nitrogen) and the hydroxyl molecule (one atom of hydrogen plus one atom of oxygen), were detected in the gaseous clouds that lie in deep space. Spurred on by this result, a group of scientists led by Charles Townes at the University of California at Berkeley set out in 1968 to search for ammonia in space. They soon found it, in the tenuous clouds of gas towards the centre of our galaxy. As a bonus they also found water there. Shortly afterwards another team at the National Radio Astronomy Observatory in West Virginia detected formaldehyde throughout our galaxy, and later in many other galaxies. Since then about one hundred organic molecules have been detected, including all those necessary for the evolution of life.

Two British astronomers, Fred Hoyle and Chandra Wichramasinghe, have recently suggested in their book *Lifecloud* that on the tiny dust grains in the interstellar clouds the conditions may well have been conducive to the formation of these chemicals into all the basic building blocks of life. If so, the seeds of life may have been universally available, cast upon the interstellar winds waiting to fall on fertile planets as soon as they were formed.

Such hypotheses do not necessarily mean that life did not originate in the primeval soup; rather they show that these basic components of life can come together in a variety of different ways – a cosmic policy of life ensurance. Furthermore it suggests that life is a widespread phenomenon throughout the Universe, a natural consequence of chemical evolution.

THE EVOLUTION OF LIFE ON EARTH

The early oceans and rock pools of Earth were probably the setting for the next step in the evolutionary drama. Here the conditions were right for these chemicals – whether they originated on Earth or in interstellar space – to come together to form larger molecules such as amino acids, enzymes and proteins. Over time these linked together into groups and chains of increasing complexity. Those which were more stable survived longer and combined with others to form yet larger units (macromolecules, as they are often called), some of them containing thousands of the basic building blocks, and millions of atoms.

Some of these giant molecules developed the ability to 'recognise' other smaller molecules. This was possible because each type of molecule had a specific three-dimensional shape; if a smaller molecule had a shape that fitted snugly into a nook in the more complex macromolecule, it could be 'recognised' much as a key fits (or is recognised by) a lock. With this ability some macromolecules (in particular deoxyribonucleic acid molecules, otherwise known as DNA) were able to arrange other smaller molecules into specific sequences. By building up sequences that were exact copies of themselves, they achieved the essence of reproduction.

Once complex self-replicating organic molecules had established a stable hold, they began forming loose associations with other complex macromolecules. More and more molecules joined the groups, until eventually they reached a stage where the groups became integrated units. In this way the simplest cells were born, some 3.5 billion years ago.

These embryonic cells could not have survived for long in the relatively chaotic environment in which they found themselves. Life probably emerged many times, only to be swallowed up again almost at once – much as any atomic nuclei formed in the primordial fireball would have been instantly annihilated by the superhot temperature. Over time the process of repeated emergence and dissolution of life would have gradually built up a more hospitable environment, until a threshold was reached, beyond which the rate of creation exceeded the rate of loss. Life would then have established a stable foothold.

The first simple cells were algae and bacteria. They did not breathe oxygen; on the contrary, they produced it through the process of photosynthesis, and cast it away as a waste product. This oxygen became the first major pollutant on the planet, for, to the the organisms of the time, oxygen was as poisonous as chlorine is to us. Initially, the free oxygen combined with minerals such as iron to produce various oxides, and so long as it was absorbed in this way, life remained safe.

After about a billion years, however, all the available iron had been turned to rust, and oxygen began to accumulate in the atmosphere. About the same time ultra-violet light, which earlier had been so valuable in the synthesis of amino acids, and thus necessary for the evolution of life, now threatened to destroy the bacteria which had evolved. Fortunately, this planetary crisis was averted. The extra oxygen combined to form a layer of ozone in the upper atmosphere, thereby preventing much of the ultra-violet light from reaching the planet's surface. Ingenuity triumphed, as James Lovelock points out, '[not] in the human way by restoring the old order, but in the flexible Gaian way, by adapting to change and converting a murderous intruder into a powerful friend.'

As the oxygen continued to build up in the atmosphere, bacteria evolved that could tolerate the 'poison'. Later, about two billion years ago, other bacteria emerged that could actually use the oxygen to extract more energy from their food than they could by simple photosynthesis. These bacteria ultimately went on to become animals, while the photosynthesising bacteria responsible for the crisis became plants.

Oxygen continued to accumulate until about 1.5 billion years ago, when it reached a concentration of 21 percent – a level which, as we saw in the previous chapter, happens to be the optimum balance between metabolic efficiency and fire risk. Thereafter it abruptly stopped increasing and has remained remarkably stable ever since.

FROM CELL TO ORGANISM

Once the oxygen concentration had stabilized at this critical level, there came a series of important steps forward in the evolutionary process:

- Some simple cells began to be integrated within other simple cells, giving birth to more complex cells. These new cells were the first to possess a well-defined nucleus, within which the genetic material was encapsulated. The development of a nucleus made it possible for two cells to come together and produce offspring containing a combination of their genetic material, i.e. sexual reproduction became possible. This opened greater opportunities for success and failure – as far as evolution is concerned, one success is worth a million failures – and allowed new adaptations to spread more quickly through a population, thus speeding up the rate of evolution.

- As a greater variety of cells evolved, some cells appeared that were able to feed off other lifeforms. (Until now cells had consumed gases, minerals, organic molecules and light-energy.) With this development, cells no longer had to build up all their complex macromolecules from scratch, but could take in many of them ready formed, as amino acids, proteins, vitamins, etc. In effect, cells were now able to consume more highly organised matter.

- The next major evolutionary development occurred around one billion years ago, as a result of a food crisis. Beyond a certain size, the cells could no longer take in enough nourishment to feed themselves. (As a cell grows, its volume increases faster than its surface area, and it is the surface area which limits the amount of food that can be absorbed.) Evolution's response was to keep the cells the same size, and to produce larger systems by clumping together. So single cells began to gather themselves into small groups and colonies, giving rise to the first multi-cellular organisms – simple sponges and later jellyfish.

- Within these communities it became more efficient for different cells to specialise in different functions. Some took on digestive tasks, some became a protective casing, and others conveyed messages to different parts of the organism. This conferred an added adaptability and stability on the organism, helping it survive greater variations in environmental conditions.

- Another important characteristic of multi-cellular organisms was that individual cells could be replaced as they died, giving the organism as a whole the ability to live far longer than its constituents.

By about 600 million years ago, more complex multi-cellular organisms such as molluscs and simple worms began to develop. Over time, these multi-cellular organisms became progressively more organised. Different cells took on more

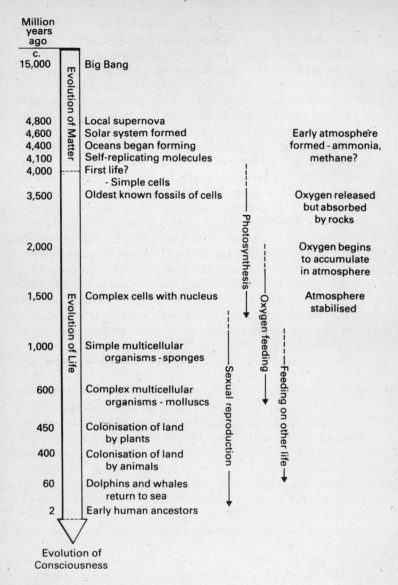

FIGURE 1. Major stages in the evolution of life.
(N.B. The scale is not linear.)

specific functions and grouped themselves into specialized organs leading to yet more complex organisms, some of them recognisable as the ancestors of modern plants and animals. About 450 million years ago, the plants began to colonise the land, and animals joined them 50 million years later. (The plants were first since they could harvest the sun's energy directly. Once they had started the food chain, the animals followed.)

This is where the evolutionary tale with which most people are familiar comes in: the story of how these early organisms evolved through numerous stages into the millions of species of plants and animals living on the planet today. Since we are going to be concerned with the major leaps in evolution rather than its detailed development, I will not go into specifics here.

The most important general trend within this part of evolution was the development of the nervous system, permitting more rapid communication between the different parts of the body. Within the vertebrates (i.e., animals with backbones) the main nerve fibres became enclosed in a protective tube (the spine), and the principal nerve centres at the top took on greater importance and became the first simple brains.

Within the last 50 million years, the brain has undergone an explosive growth – one of the most rapid and dramatic changes in the history of evolution. If we look at the ratio of brain weight to body weight, and start the scale with earthworms and insects as 1, then stenonychosaurus (one of the most intelligent dinosaurs, living about 75 million years ago) was around the 20 mark, while human beings have leapt up to the 350 mark.

Even more significant, the cortex – the outer layer of the brain thought to be the seat of the higher mental functions – has become relatively thicker and larger. The most developed cortices on this planet are to be found in human beings and in some of the cetaceans (dolphins and whales). Whether dolphins and whales are more or less intelligent than humans is an open question; research in this area is still in its infancy. Indeed, the question is probably unanswerable in that it presumes some absolute definition of intelligence that is equally applicable to humans and dolphins alike.

The brains of whales and dolphins appear to have stopped evolving 20 million years ago – long before the first humans appeared – suggesting that they may be perfectly adapted to their watery environment. The human brain, on the other hand, is a relatively new evolutionary venture, having evolved over the last 3 million years or so. It is almost certainly still evolving – though our human timeframes may prevent us from being directly aware of this.

SELF-REFLECTIVE CONSCIOUSNESS

With the development of the large human brain and cortex another major evolutionary leap occurred, as significant as the emergence of life. This was the emergence of self-reflective consciousness. Humans are not only conscious; they are conscious of being conscious.

Consciousness is a very difficult subject to be clear about. For a start, the word 'consciousness' has been used in many different ways by different people. One dictionary definition of the word is 'knowing of external circumstances', which would imply that being asleep is a state of unconsciousness. Yet we certainly have experiences when we dream; we are still conscious internally. Moreover, a person in a coma or under anaesthesia may appear completely insensible to the people around, yet afterwards may report having experienced what was happening. Another use of the word relates to the amount of attention put into an experience or action. If, while eating a meal, we are more engaged in the conversation than the meal, we might be said to be eating without consciousness of the process. Or, we may use it in the sense of intent or deliberation, as in making a choice with full consciousness of its consequences. We also talk of a person's social, political or ecological consciousness, meaning the particular way he looks at and evaluates the world.

The difficulty surrounding the meaning of the word 'consciousness' arises in part from the fact that in English we have only one word to convey so many different meanings. In Sanskrit, the ancient Indian language, there are some twenty different words for consciousness, each with its own specific meaning, some representing concepts with which we in the West are barely familiar. (*Chitta*, for example, is the 'mind-stuff' or 'experiencing medium' of the individual; *chit* is the 'eternal consciousness' of which the individual mindstuff is a manifestation; *turiya* is the experience of pure consciousness without an object; *dhyana* is consciousness focussed on an idea; *purusha*, the essence of consciousness, is somewhat akin to the Holy Spirit.)

Here I shall be using the word 'consciousness' to mean the 'field' within which all experience takes place. In this sense consciousness is a prerequisite for all experience whether we are awake, in a trance, dreaming, in a coma, or in any other state.

Used in this sense, consciousness is not restricted to human beings; any being which experiences has consciousness. Anyone who has spent time with other mammals, such as dogs, cats or horses, has probably come to the conclusion that they are also conscious beings. They 'know' what's going on. They are not automata. Birds, reptiles and fish would also appear to have consciousness, and maybe insects, snails and worms do as well. According to some researchers, even plants appear to have some type of awareness.

An important attribute of conscious beings is the ability to form internal models of the world they experience; the greater the consciousness, the more complex the models. A worm probably has a relatively simple model of reality, whereas a dog's model would be considerably more complex. In human beings the nervous system has evolved to a point where our internal models of reality are so complex that they have had to include the self – the 'modeller' – in the model. This is the beginning of self-reflective consciousness. We not only experience the world around and within us, we also are conscious of ourselves in that world, and conscious that we are conscious.

The emergence of self-reflective consciousness is to some degree tied in with the development of verbal language. Language allowed us to communicate more widely and more fully. It also allowed us to focus attention on abstract and even hypothetical qualities of our experience. With this tool we could even begin to consider the nature of our own experience, enabling us to separate the 'experienced' from the 'experiencer' (the self) – a separation and objectification which, as we shall see in Chapter Seven, has its drawbacks as well as its advantages.

The development of language was also significant in that it led to the exchange of information between individuals. Thus a person could gain from the successes and failures of others, rather than have to learn everything from scratch. Humanity has also developed writing, which is the ability to store language in symbolic form. With writing came the ability to transfer information across time. This was as significant for the speeding of evolution as was the development of sexual reproduction – also an information transfer – in the single cell. The later invention of printing, and the more recent developments of photocopying and telecommunications have likewise played a major part in accelerating the evolution of civilisation.

So our brief review of evolution on this planet brings us to the present day. Suddenly, in a flash of evolutionary time, a new species has emerged: one that is aware of its own existence, and one that holds awesome potential for consciously affecting itself and its environment.

This almost unbelievable product of 15 billion years of evolution is truly something to be marvelled at. Here we are, each of us several septillion atoms, arranged into an integrated system of some hundred trillion biological cells, experiencing the world around us and the thoughts inside, experiencing various emotions and various desires. And we are, above all, conscious of all these things, and conscious of being ourselves. We can communicate these experiences to others, in words and in a variety of other ways. We can imagine alternative futures, and make choices to bring them about. We can even imagine the impossible. Furthermore, we can sit here and wonder at the whole evolutionary process which has step-by-step resulted in me and in you, in farms, automobiles and computers, in men walking on the moon, in the Taj Mahal, the Emperor Concerto and the Theory of Relativity.

If anyone had been around 4 billion years ago, would they ever have guessed that the volcanic landscape, the primeval oceans, and the strange mixture of gases in the atmosphere would steadily evolve into such an improbable and such a complex being? If told, would they have believed it?

Could we now, if we were told what would happen in the next 4 billion years of evolution, believe it? Would the future seem as improbable to us as we were at the birth of the Earth? What unimaginable developments lie ahead, not only in thousands of millions of years' time, but in just one million years?

And what of the next few thousand years? The next hundred even? Where is evolution likely to take us next? To see, we need to look at some of the general trends and patterns which have characterised the whole evolutionary process thus far.

CHAPTER 3

Hidden orders in evolution

Driven by the force of love
the fragments of the world
seek each other that the world
may come into being.

Teilhard de Chardin

When considering the vast panorama of evolution, it is easy to get lost in the many changes and developments that have taken place. But if we stand back and consider the evolutionary process as a whole, a number of patterns begin to become clear.

One of the first things we notice is that the Universe today contains characteristics of which there was no trace in the beginning. Immediately after the Big Bang, there was only *energy*. Out of this developed a whole new order of existence: physical *matter*. For eons, this matter was inanimate, yet out of it emerged a new order: *life*. Life persisted and flourished and from living organisms emerged another new order: self-reflective *consciousness*.

Each of these new orders of existence represented a major step forward in the evolutionary process, bringing with it novel properties and characteristics which could not be predicted from the previous stages. The new whole became more than the sum of its parts, and not predictable in terms of its constituents.

We can see this happening with the progression from energy to matter, life and consciousness. Pure mathematics, for example, which is sufficient to describe the electromagnetic radiation of energy, does not readily predict the behaviour of molecules; this is the realm of chemistry. Chemistry likewise does not predict the principles that govern living organisms. And biology cannot readily describe conscious experience. Each of these levels are new phenomena – new emergent orders of existence.

This is not to say that the laws of the lower levels are no longer valid. The elementary particles in a cell continue to obey the laws of physics, the atoms continue to obey the laws of chemistry, and the macromolecules perform as expected by molecular biologists. Each new order subsumes all the previous orders; nothing is lost. Yet, something new is created, and the new phenomenon brings patterns of behaviour that require a new level of understanding and explanation.

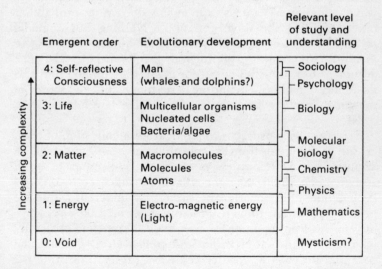

Emergent order	Evolutionary development	Relevant level of study and understanding
4: Self-reflective Consciousness	Man (whales and dolphins?)	Sociology / Psychology
3: Life	Multicellular organisms Nucleated cells Bacteria/algae	Biology
2: Matter	Macromolecules Molecules Atoms	Molecular biology / Chemistry / Physics
1: Energy	Electro-magnetic energy (Light)	Mathematics
0: Void		Mysticism?

(Increasing complexity →)

TABLE 2. Emergent orders of evolution showing fields of study relevant to the different stages. I have termed the state of affairs before the Big Bang the zeroth level, and called it the 'Void'.

Western science sometimes finds it difficult to deal with the notion of emergent orders of existence. This is because one of the principal ways in which it has tried to understand the world is to break phenomena and processes down into smaller units, what is called the reductionist approach. Although valuable in some areas (such as physical chemistry, engineering and computer programming), it has the drawback that emergent qualities of the system as a whole are usually lost or not dealt with.

The reductionist approach argues that consciousness can be explained in terms of neural events in the brain, and life in terms of organic chemistry. Taken to its logical conclusion, this argument ends up in a trap of its own making. Consciousness, it is said, is 'nothing but' the cumulative effect of a complex interwoven web of 10 billion nerve cells. A nerve cell is 'nothing but' a huge conglomeration of macromolecules. A macromolecule is 'nothing but' a few million atoms strung together, and an atom is 'nothing but' a nucleus surrounded by a cloud of spinning electrons, which in turn are 'nothing but' eigenvalues in a probability function called the wave equation. What is 'an eigenvalue in a probability function called the wave equation'? Nothing but a model created by the conscious processes of the human mind to give meaning to certain experimental results in physics. The argument has come full circle, for is not the human mind and its many faculties, including creativity and a sense of meaning, 'nothing but' the workings of a few billion brain cells?

Clearly consciousness is different from a collection of cells, as life is different from a collection of atoms. Instead of arguing that consciousness is merely a by-product of brain activity, one could take the view that since consciousness evolves out of life, consciousness is already inherent within life in some potential, though unmanifest, form. Likewise, since life evolves from apparently inanimate matter, life is already inherent within matter in an unmanifest form. Perhaps the potential for each new order is always present, awaiting the particular conditions that would allow it to manifest.

What are these conditions? The answer would seem to lie partly in a progressive increase in complexity.

EVOLVING COMPLEXITY

The word *complex* does not simply mean 'having many parts'. It also implies that the many parts are interrelated and independent (the Latin root of complex means interwoven or plaited). In contemporary systems theory complexity is sometimes defined as a measure of the number of relationships between the different components of a system. The concept of complexity might therefore be thought of in terms of three basic characteristics:

diversity – the system contains a large number of components, usually of various different kinds
organisation – the many components are organised into various interrelated structures
connectivity – the components are connected through physical links, energy interchanges, or some form of communication. Such connectivity maintains and creates relationships, and organises activity within the system.

In other words, for something to be called complex, it must be composed of a number of different elements which are organized in some way, and connected so as to interact with one another. To get a clearer idea of what this means, let us look briefly at these characteristics at work in evolution.

DIVERSITY

The bacterium *Escherichia coli*, which lives in the human intestine, is one of the simplest forms of life. Yet one such cell contains four DNA molecules, (each containing several hundred million atoms), about 400,000 RNA molecules (of 1,000 different kinds, and each containing a 100,000 or so atoms), about one million protein molecules, (of 2,000 different kinds, and each with an average of a thousand atoms), and some 500 million smaller organic molecules.

Diversity is also very apparent in a complex organism such as the human being which contains many different types of cell (e.g. liver cells, brain cells, skin cells, blood cells, bone cells and so on).

As far as the emergence of new evolutionary levels is concerned, a critical number of basic components seem to be required. The basic components of a living cell are atoms (stable units of matter). Within each cell of *Escherichia coli* there are about 40,000,000,000 atoms (written in mathematical shorthand as 4×10^{10}, where 10^{10} means ten raised to the tenth power, i.e. '1' followed by ten zeroes). More complex cells such as a muscle cell may contain 10^{12} atoms, and some large amoeba may contain as many as 10^{15} atoms. In the other direction, however, we can find only a few types of cell containing less than 10^{10} atoms, and there are no known forms of life with less than 10^8 (i.e. a hundred million atoms). In terms of sheer numbers, there would seem to be a threshold below which life does not readily emerge.

A similar threshold appears to exist for the emergence of self-reflective consciousness from life. The average human brain contains about 10^{11} nerve cells, of which 10^{10} are in the cortex, the area associated with conscious thought processes. Brains with cortices containing 10^9 or fewer neurons, such as the brain of a dog, do not appear to show the phenomenon of self-reflective consciousness. Only when you reach the size of the human cortex does this faculty emerge, and with it the development of thinking, language, the intellect, knowledge, free will, science, art and religious experience.

We might therefore postulate that something of the order of 10^{10} units are required before a degree of complexity can be established which is sufficient for a new order of existence to emerge. If the number of elements collected together is significantly less than this number, there is insufficient scope for the organisation and interrelationships which are also necessary.

ORGANISATION

A macromolecule, such as a protein, is no mere random aggregation of atoms. Rather, the atoms are organised in a very specific structure, and just one atom out of place can radically change the characteristics and properties of the molecule. The millions of macromolecues that constitute a living cell are themselves organised into various organelles (the 'organs' of the cell), which are in turn organised in specific ways, interact with each other at specific times for specific purposes, and generally synchronise their activities in ways that biologists are only just beginning to comprehend.

Similarly our own bodies are highly organised collections of cells and organs. (The words organism and organize, as you might have suspected, are derived from the same root.) Each of us contains some 10^{12} living cells, totaling around 5×10^{25} atoms, all organised in a very particular way so that the end result is a human being rather than a large bowl of soup.

CONNECTIVITY

Not only is the structure of a complex organism organised, but so are its internal activities. The various parts are connected together physically, they exchange matter and energy, and there is a flow of information between the many components and subsystems. The information aspect is going to be crucial in later discussions, so let us look briefly at how it manifests at various levels.

At the level of elementary physical matter, particles exchange rudimentary information about their nature (e.g. their charge or spin) and location through the basic physical forces (the gravitational force, the electromagnetic force and the weak and strong nuclear forces.) If, for example, two particles experience an electrostatic repulsion, they could be said to be receiving information that they are of opposite charges.

Further up the evolutionary scale, complex macromolecules exchange information about their shape and structure, as some molecules fit into the contours of other molecules like pieces in a jig-saw puzzle.

In simple living cells, information is transferred through the duplication process of asexual reproduction. The further evolutionary advance of sexual reproduction enabled the genes to transfer information from two parent cells to a new cell, thus increasing the quantity of information flowing from one generation to the next.

In simple organisms, information flows mainly with the aid of chemical messengers: hormones provide communication

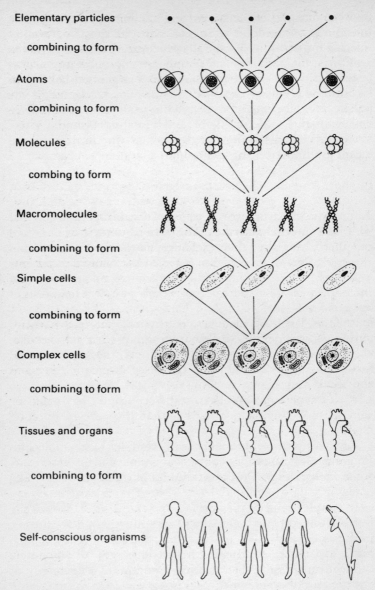

Elementary particles

combining to form

Atoms

combining to form

Molecules

combing to form

Macromolecules

combining to form

Simple cells

combining to form

Complex cells

combining to form

Tissues and organs

combining to form

Self-conscious organisms

FIGURE 2. Evolution as a progressive collecting together of units into larger systems.

between one part of the system and another, and pheromones (chemicals released into the environment) convey information to other organisms (e.g., the airborne sex attractants of moths and the scent trail of ants.) Within more complex organisms, faster and more versatile forms of communication have been incorporated employing the transmission of electrical impulses along nerves. Moving up to human society, we find the development of a variety of forms of interpersonal communication, such as language, writing, art and music, and the recent addition of electronic communication networks.

In short it would appear that one of the principal trends of evolution is a movement towards increasing complexity: individual units gather together into larger and larger groups, which display increasing organisation and structure as they expand, the many components interrelating in various ways.

Yet complexity appears to be more than just an evolutionary trend. It also seems to be the prerequisite for the emergence of new levels of evolution. Only when energy becomes organised in a particular way can the qualities of matter emerge and manifest themselves; only when many units of matter become collectively organised in a particular way can life emerge and manifest itself; and only when many living cells become collectively organised in a particular way does consciousness emerge and manifest itself.

ORDER vs. DISORDER

This evolutionary trend towards increasing complexity and organization might appear to be in opposition to a well-established law of physics – the second law of thermodynamics – which implies that the Universe as a whole is moving steadily towards increasing disorder.

The second law of thermodynamics states that in any energy interaction there is always a reduction in the amount of energy available to perform useful work. If, for example, you burn a piece of wood, energy is transformed from the energy of various chemical bonds into heat energy. Some of this energy could be harnessed to do useful work; it might heat the boiler of a steam engine, for instance. However, we can never again burn that piece of wood; the amount of free energy available has decreased. Nor can you get back to the original state by mixing the ashes, smoke and heat together.

Physicists measure the amount of energy in a system which is no longer available to do useful work in terms of a concept called entropy. When the available or free energy decreases, the entropy is said to increase.

Entropy is also a measure of the amount of randomness in a system. When the entropy is at a minimum, the internal order of a system is at a maximum. As the entropy increases, the system becomes more disordered. Thus the second law of thermodynamics also implies that after any energy exchange, there is an increase in the disorder in the system.

As a simple example, imagine a drop of ink in a bowl of water. As the ink spreads out, its molecules will go from a more concentrated and organised state to a more random distribution. To the non-physicist it might seem that the evenly distributed mix is the most orderly, but to a physicist or mathematician the most orderly state is the one in which the positions of the ink molecules are most easily defined, which is when all the ink molecules are localised in a single drop. As the ink spreads and the molecules become more randomly distributed, this mathematical order decreases and the entropy increases.

An important consequence of the second law is that such processes are irreversible. Systems do not spontaneously become more ordered; left on its own an ink solution does not return to being a concentrated drop. Since this law applies to all physical systems, the entropy of the Universe as a whole must also be increasing. In other words, the physical Universe is continually running down, moving towards more random, less orderly states.

Life, however, would appear to contradict this trend. Living systems display a great deal of order. Every living being, from *Escherichia coli* to a blue whale, is a highly organised collection of energy and matter. And over time individual living systems not only retain a high degree of internal organisation, they build up this order as they grow and develop. Life appears to move towards increasing order rather than disorder. But according to the second law of thermodynamics, a system such as your body should be gaining entropy, returning to a pre-biological soup. So does life somehow contravene a well-established and seemingly universal law of physics?

The answer is 'No'; and the reason is that the second law only applies to *closed* systems – systems that are isolated from their environment such that there is no flow of matter or energy in or out of the system. (An example of an ideal closed system would be a sealed container, perfectly insulated and impervious to vibration, sound, light, magnetic fields, X-rays and any other form of energy transmission.)

Living systems, however, are *open* systems, continually exchanging matter and energy with their environment. When we consider an organism *plus* the whole of its environment as a single system, the second law of thermodynamics still holds because we are now effectively considering a closed system, and the overall entropy increases. Bacteria living inside a closed container, for example, will show a decrease in entropy (i.e. an increase in their internal order), although the total entropy of the container and bacteria *together* will have increased.

Considered on its own an organism effectively retains internal order at the expense of order in the environment. In the words of the great physicist Erwin Schrodinger: 'What an organism feeds upon is negative entropy; it continues to suck orderliness from its environment.' Or, to put it another way, an organism exports entropy to its environment. This export may occur through the excretion of less-ordered material or through the emission of heat (heat energy being of high entropy). The net effect is that the resulting local decreases in entropy associated with a living system are paid for by larger increases in the entropy of the environment.

Nevertheless, even though living processes may not contradict the second law of thermodynamics, the question must be asked as to *why* an organism builds up and preserves a high degree of internal order. Why does a certain collection of atoms go against the trend of the rest of the Universe? Indeed, if the entire evolutionary process can be seen as one of increasing organisation, why does this happen within a Universe that is, as a whole, running down towards disorder?

SELF-ORGANISING SYSTEMS

For a long while there were no satisfactory answers to these questions. In the 1970s, however, Ilya Prigogine, a Belgian physical chemist, working in Brussels and at the University of Austin in Texas, made a major breakthrough in understanding how order can arise from disorder – a breakthrough for which he won the 1977 Nobel prize in chemistry. He noticed that there were a few physical and chemical systems that could build and maintain a high degree of order in their physical structure even though no such order was fed into them.

One particular chemical reaction, known as the Belousov-Zhabotinsky reaction, was the subject of considerable investigation by Prigogine and his colleagues, since it provided an excellent example of patterns of organisation arising from a homogenous mixture of substances. In this reaction, four particular chemicals (malonic acid, a sulphate of cerium, potassium bromate and sulphuric acid) are mixed up in specific concentrations and left in a shallow dish to react. Within a few minutes concentric or spiralling waves are seen spreading out across the dish, and these patterns may continue for several hours.

An important characteristic of the chemical process displayed here is that the reactions are cross-catalytic, i.e., the products of one stage act as catalysts for later stages. As a result they go through a series of repetitions, and it is this which gives rise to the distinctive patterns. The appearance of these ordered patterns represents a decrease in entropy within the dish, made possible by the export of an even greater amount of entropy to the surroundings. The net entropy of the whole system (the dish plus its environment) has still increased, satisfying the second law of thermodynamics.

Prigogine termed such self-ordering processes *dissipative structures*, since the entropy they produce is dissipated to the environment. A dissipative structure always produces entropy, but gets rid of this entropy through its continuous interaction with the environment. As energy and matter are taken in, entropy (usually in the form of heat) and some end products are expelled – a process which we might justifiably refer to as the 'metabolism' of the system.

Through their work, Prigogine and his colleagues have found that three conditions are necessary before a dissipative structure can form:

Openness: Matter and energy must be able to flow between the system and its environment.
Far from equilibrium: Only if the system is far from the state of thermodynamic equilibrium can self-organisation persist. Near equilibrium the system behaves like any other physical system – there is increasing entropy.
Self-reinforcement: Certain elements of the system catalyse the production of new elements of the same kind, i.e. the elements are self-reproducing.

If the energy or matter flowing through a dissipative system fluctuates, the internal organisation will be retained as long as the fluctuations remain within certain limits. The system can even suffer minor physical damage, yet 'heal' itself through its self-organising nature. If, however, the fluctuations increase beyond a certain limit, they will ·drive the system into instability. From this state the system may collapse; alternatively, it may make a transition to a new level of organisation. In other words, a dissipative system is capable of *evolving* in response to major fluctuations in the environment.

Characteristic of these transitions is a period of considerable chaos within the system. This instability corresponds to a maximum flow of energy and matter through the system, and maximum production of entropy within the system, requiring maximum dissipation of entropy into the environment. Should the system survive this period, the result can be a reorganisation and a new regime of dynamic stability. Such a phenomenon has important implications for evolution in general.

DISSIPATIVE STRUCTURES IN EVOLUTION

The general behaviour of dissipative structures is clearly very similar to that of living systems. And this is not merely a superficial similarity. Prigogine has shown that biological processes, far from being renegades in an otherwise disordering Universe, are actually predictable by the principles of dissipative structures. He concludes: 'Life no longer appears as an island of resistance against the Second law of Thermodynamics . . . it would appear now as a consequence of the general laws of physics . . . which permit the flow of energy and matter to build and maintain functional and structural order in open systems.'

Biological systems have now become the prime focus of research in this field. Phenomena such as the growth of plants, the regeneration of limbs in simple organisms, the excitation pattern of nerve cells, and many biochemical processes are now well understood as extensions of the principles found in dissipative structures. This theory has been expanded to include the study of the brain, and has also been applied to social groups such as bee swarms and slime moulds, to human economic interactions, to human society, to ecosystems, and even to Gaia herself. As a result, the theory of dissipative structures has brought some major advances in our understanding of how living systems develop.

The theory can also be applied to evolution in general. Eric Jantsch has explored this subject comprehensively in his book *The Self-Organising Universe*, showing that the continual trend of evolution towards increasing complexity can, at each stage, be explained as the effect of dissipative systems.

As mentioned above, extreme fluctuations within dissipative structures can lead to the emergence of new levels of organisation. In evolutionary terms these fluctuations appear as periods of instability, or crises, in which organisms are forced either to adapt to the changed environment – perhaps moving on to higher levels of organisations – or be extinguished.

An early crisis (or fluctuation) in the evolution of life possibily occurred when the simple organic compounds on which the first primitive cells fed started running short. This was, in effect, the first food crisis. The response was the evolution of photosynthesis – the ability to feed directly from sunlight. Photosynthesis produced oxygen as a by-product, and one-and-a-half billion years later, as the oxygen began to build up in the atmosphere, there was another major change in the environment and another crisis, this time of pollution and poison. The response was the evolution of oxygen breathing cells. Later, as cells became more complex and grew in size they faced another crisis – they could not absorb food fast enough to nourish themselves. This time the evolutionary response was to develop multicellular organisms.

In the present day it is readily apparent that society is also going through some major crises. Looking at humanity from the perspective of dissipative systems, we can see that the two principal characteristics of a major fluctuation are present: increasing throughflow of energy and matter, combined with high entropy. We are now consuming energy and matter like never before, with all the ensuing problems of resource scarcity and depletion. At the same time the entropy produced by humanity has shot up. This is apparent as increasing disorder both within society (e.g. rising social unrest, increased crime and growing economic chaos), and in the surroundings (e.g. increasing despoliation of the environment and rising pollution.)

We would seem to be rapidly approaching the breaking point. And there are two possible outcomes – breakdown or breakthrough. If we cannot adapt to the pressures being brought to bear, human society will probably collapse. If we can adapt we may move on to a new level of organisation. Which path we take is now very much up to us. One thing seems clear: the pace of change is increasing, and whichever route we take major changes are not far away.

CHAPTER 4

The increasing tempo of evolution

It is commonplace today to speak of the pace of life speeding up, and to look back with nostalgia at the more leisurely pace of our grandparents' time. But this speeding up is not new; it has been going on for the last 15 billion years. In evolution each new development was able to draw upon what had already been accomplished. (The evolution of complex macromolecules, for example, was able to utilize the properties and characteristics of less complex molecules such as amino acids, aldehydes and water.) Each new phenomenon became another platform that evolution could use in its relentless movement towards yet greater complexity. The more that had already been accomplished, the broader the platform on which evolution could then build, and the greater the rate of development. This produced accelerating patterns of growth.

As a result of this natural tendency to accelerate, the major developments in evolution have not occurred at regular intervals; rather, the intervals have been shortening. When we talk in terms of billions of years, however, it may sometimes be difficult to see this happening. Such concepts are way beyond our experience. To get a more tangible image let us compress these 15 billion years into a film a year long – the ultimate 'epic'. And so that you may fully appreciate the timescales involved, imagine what it really would feel like sitting through this film.

The Big Bang, with which the film opens, is over in a

hundred millionth of a second. The Universe cools rapidly and within about 25 minutes stable atoms have formed. No more significant changes happen during the rest of the first day, nor for the rest of January (you will need plenty of popcorn): all that you are viewing is an expanding cloud of gas. Around February and March the gas clouds begin slowly condensing into clusters of galaxies and stars. As the weeks and months pass by, stars occasionally explode in supernovae, new stars condensing from the debris. Our own sun and solar system are eventually formed in early September – after eight months of film.

Once the Earth has formed, things begin to move a little faster as complex molecules start to take shape. Within 2 weeks, by the beginning of October, simple algae and bacteria appear. Then comes a relative lull (and more popcorn) while bacteria slowly evolve, developing photosynthesis a week later, with the consequent build up of an oxygen atmosphere after 5 more weeks, in early November. Within another week, complex cells with well-defined nuclei evolve, making sexual reproduction possible, and with this stage accomplished, evolution accelerates again. It is now late November, and the major part of the film has been seen. The evolution of life, however, has only just begun.

The first simple multi-cellular organisms appear around early December, and the first vertebrates crawl out of the sea onto the land a week or so later. Dinosaurs rule the land for most of the last week of the film, from Christmas to mid-day on December 30th – a long and noble reign!

Our early apelike ancestors made their debut around the middle of the last day, but not until 11 o'clock in the evening do they walk upright.

Now, after 365 days and nights of film, we come to some of the most fascinating developments. Human language begins to develop one-and-a-half minutes before midnight. In the last half minute farming begins. Buddha achieves enlightenment under the bodhi tree five-and-a-half seconds before the end, and Christ appears a second later. The Industrial Revolution occurs in the last half second, and World War II occurs less than a tenth of a second before midnight.

We are down to the last frame now, the last inch of a

hundred thousand miles of film. The rest of modern history happens in a flash, not much longer than the flash with which the film started. Moreover, evolution is continuing to accelerate, and this rapid acceleration shows no signs of abating.

EVOLUTIONARY LEAPS

If we plot the evolutionary changes we have just reviewed, we would see that within the acceleration were periods of rapid growth followed by periods of much slower development. Acceleration has not occurred smoothly: rather, it appears to have happened in a series of sudden steps.

Let us consider, for instance, the process that occurred after the sudden formation of stable atoms of hydrogen, some 70,000 years after the Big Bang. Over billions of years, these atoms slowly evolved into other simple elements. The greater the variety of atoms that were created, the greater the potential for more atoms to form. Thus, over time, this platform-building process accelerated until a stage was reached where most of the heavier elements were formed quite suddenly, in a period of some 15 minutes, in the heart of exploding stars.

Another series of accelerating steps occurred in the transition from matter to life. Before the development of stars with planetary systems, the only mechanisms open to evolution were at the atomic and subatomic levels. Once the cooler conditions necessary for the formation of more complex molecules existed, evolution had another foothold on which to build and could go that much faster. The more complex the molecules it produced, the greater scope there was for further evolution. So the process speeded up until simple bacteria emerged.

This occurred relatively early in the Earth's history, some 4 billion years ago. At this stage evolution stepped onto a new platform, the platform of life. But life at this time was still very simple and offered little variety, and the process slowed down again. It took between 10 and 20 times as long for cells with a simple nucleus to emerge from bacteria as it did for bacteria to emerge from the prebiotic chemical soup. But once nucleated cells appeared they were capable of sexual reproduction, which led in turn to greater variation and adaptability. With the continued creation of more and more complex forms, the platform of life expanded and evolution moved on faster and faster.

When we come to consider the evolution of the many different animal species, it appears that this too was probably an intermittent process. The classical view of biological evolution, based on Darwin's theories, supposed that a species evolved slowly through a long series of minor changes, and over millions of years this process led to the emergence of a completely new species. While most of the general principles of Darwinian evolution are accepted by the vast majority of scientists, the notion of continuous gradual development has recently been called into question.

We ought, according to this view, find fossils representing a smooth progression from one species to another; instead, we find a large number of fossils for one species, and a large number for the species that it appears to have evolved into, but very little in between. There are too many missing links.

Some evolutionary theorists, such as Stephen Jay Gould from Harvard, now believe that individual species have enjoyed long periods of stability followed by periods of rapid evolution. Just how quickly these evolutionary leaps take place is still unclear; proponents of this view suggest they may happen in 50,000 years or so (which in evolutionary terms is quick compared to ten million years), although they could, in the right circumstances, occur in only a thousand years, and there are even cases in the present day of species of birds and moths have been observed undergoing major evolutionary changes in just one generation. Such findings suggest that the evolution of life was not nearly as smooth as Darwin originally proposed.

With the appearance of humanity, evolution moved on from the biological level to a new level – consciousness. We are, it is almost certainly true, still evolving as a species, but this process, rapid as it may be from an evolutionary perspective, is occurring relatively slowly as far as human timescales are concerned. As far as we can tell, we are physiologically very similar to human beings ten thousand years ago. What is evolving, and evolving very rapidly, is the human mind and the ways in which we apply it.

Self-reflective consciousness brought with it the ability to direct our own destiny. We are not bound to a long slow process of evolution through trial and error, and the selection of the correct adaptations; we can anticipate the results of our actions and consciously choose those which are most likely to take us where we want to go – as individuals and as a species. As a result, human evolution has taken a huge leap forward; so much so that we now appear to be in the midst of an unprecedented period of extremely rapid development.

ACCELERATION TODAY

The rate of change in many areas of activity is now so fast that it is difficult to predict where we will be in fifty year's time, let alone have any idea of civilisation in a thousand or a million

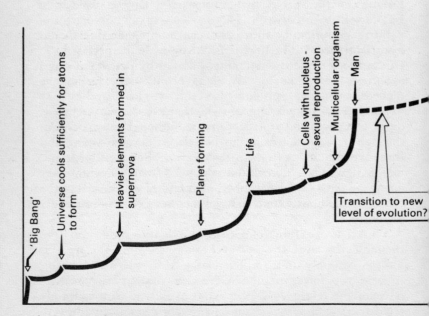

FIGURE 3. Evolution as a series of sudden transitions.

years. Moreover, many of the changes represent developments in evolution as significant as such earlier developments as sexual reproduction and photosynthesis. Never before has a product of evolution participated so actively in accelerating the evolutionary process as humanity today. Here are just a few examples:

BIOLOGY

Like any other science, this field has shown a consistent acceleration in its development. Starting with the Babylonians around 2000 B.C., and continuing with the Chinese, Egyptians, Greeks and Romans, the science grew slowly through the Middle Ages. It accelerated in the Renaissance, and even more so in the seventeenth century with the development of the microscope and the discovery of microbes. Later developments included the classification of species, the discovery of cells, and an understanding of genetic principles, all of which pushed biology ahead even faster.

In the twentieth century, the science accelerated further through the development of biochemistry, more precise instruments, electron microscopy and computer processing. Today we even know the detailed molecular structure of a gene.

But recently human beings have become more than just passive observers of the living world. Within the last decade biologists have also learnt how to modify the genes in a cell, opening the door to the creation of completely new species. No longer must the evolution of new lifeforms follow the slow process of trial and error, and natural selection. They can be consciously designed and created within a matter of months.

From an evolutionary perspective, this is a most significant event. The only innovation which previously expanded life's ability to diversify itself on such a widescale was the development of sexual reproduction by simple cells two billion years ago. Yet even this capacity took a billion years to evolve; human science has achieved a comparable step in just a few hundred years.

ATOMIC PHYSICS

Equally significant developments have occurred in atomic physics. The notion of atomic structure was first proposed 2,500 years ago by the Greek philosopher Leucippus and his follower Democritus. Twenty three centuries later, in 1808, the British chemist John Dalton discovered that different elements have different atomic weights. After the discovery of electrons in 1897, it was realised that atoms were not the smallest units but were comprised of yet smaller particles. By the 1930s physicists had settled on a standard atomic model: a nucleus consisting of protons and neutrons, with electrons in discrete orbits around it. This marked the birth of atomic physics.

With the advent of particle accelerators, scientists once again became more than just passive observers. They were now able to change some elements into others, or even create completely new elements, by bombarding the nucleus with atomic particles and thereby changing its structure.

The evolutionary significance of this development can be more fully grasped when we realise that the last time new elements were synthesised in this area of our galaxy was in a supernova explosion prior to the Earth's formation. In other words, humanity is initiating a process which has not occurred in this part of the Universe for over four billion years.

ENERGY SOURCES

Ultimately all our energy comes from the sun. (It could be argued that nuclear power is an exception since it draws on energy bound into atoms created before our sun's birth; yet this energy came initially from an earlier sun.) On Earth the major conversion of the sun's energy occurs through the photosynthesis of light by plants. The energy in wood, peat, coal and oil was all initially produced by photosynthesis, though in some cases millions of years ago. In the last hundred years we have created a fundamentally new way of directly harnessing the sun's energy – the solar cell. This invention represents an evolutionary development as significant as that of photosynthesis 3.5 billion years ago.

MOBILITY

Over the millennia human travel has progressed from walking, to horseback, to boats, to trains and cars, to supersonic jets, and to space rockets travelling at 25,000 miles per hour. At each stage there has been a leap in speed and distance coverable, and each increase has been greater than the one before. Moreover, the intervals between each new development have been rapidly shortening. We now stand on the threshold of the colonisation of space, a development as significant as the colonisation of land by the first amphibians 400 million years ago.

COMMUNICATION

Humanity's ability to transfer information has progressed from speech, to drawing, to writing, to printing, to telegraph, to telephone, to radio, to television, to photocopying, to word processing and satellite linkups. The significant increases here have been in the quantity, quality and availability of information. The combined effect of these developments has been the progressive linking up of humanity, a trend which, as we shall see later, is crucially important for the further evolution of humanity, and whose evolutionary parallel can be found in the emergence of the first multicellular organisms one billion years ago.

Were only one of these developments occurring today, we would be living in an evolutionarily important time. But the fact that these momentous changes are occurring simultaneously suggests that we are in the midst of a phase that has no evolutionary precedent.

Moreover, these developments are cross-catalytic – progress in one field accelerates the progress of another. (Molecular biology, for example, has been accelerated by microscopy, computer analysis, chemical theory, chromatography and microanalysis; the resultant advances in molecular biology have in turn fed back into areas such as medicine, agriculture, chemistry, industry and technology.) This further speeding up, through the similtaneous convergence of so many different growths, is itself something that has happened only rarely in the history of evolution, and never as rapidly nor with so many widespread and diverse implications.

It is becoming increasingly difficult to avoid the conclusion that we who are alive today are at a unique point in evolution. As John Platt, a systems theorist long fascinated by the acceleration of evolution, wrote in *The Futurist*:

Jumps by so many orders of magnitude, in so many areas, with this unprecedented coincidence of several jumps at the same time, and these unique disturbances of the planet, surely indicate that we are not passing through a smooth cyclical or acceleration process similar to those in the historical past. Anyone who is willing to admit that there have been sudden jumps in evolution or human history, such as the invention of agriculture or the Industrial Revolution, must conclude from this evidence that we are passing through another such jump far more concentrated and more intense than these, and of far greater evolutionary importance.

If this is so, if the rapid acceleration so characteristic of today is heading us towards an evolutionary leap, what lies beyond? Could we be on the threshold of a leap as significant as the evolution of life from inanimate matter?

CHAPTER 5

Our evolving society

Remember
that you are at an exceptional hour in a unique epoch,
that you have this great happiness,
this invaluable privilege,
of being present at the birth of a new world.

The Mother (Auroville)

From the birth of the Universe to the present time, evolution has inexorably pushed torward greater and greater levels of complexity: increasing diversity, organization, and connectivity. Out of this growing complexity, new orders of evolution have emerged. There is no logical reason to suppose that this trend should stop now. On the contrary, it shows every sign of continuing. The three principal aspects of complexity appear to be once again reaching the point where a new order of existence could emerge, and the arena for this next evolutionary breakthrough is humanity itself.

For us to assess the level of complexity in society today, let us look first at the matter of increasing diversity. In evolutionary terms, diversity has two major aspects. The first is variety – the development of a wide range of types within the group. Clearly this has happened within humanity. No matter how we subdivide the members of the human species (by nationality, race, body type, profession, or belief system), there is no shortage of variety.

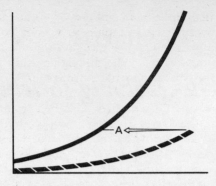

FIGURE 4. An exponential curve, typifying the growth of a population, money invested at compound interest, or any other growth in which the rate of increase is directly proportional to the current size.

Two words of caution about this curve: Firstly, since the curve gets steeper and steeper, it is sometimes mistakenly said to be asymptotic, meaning that it would eventually be going up vertically. But this never actually happens; if it did it would imply that at a certain time the population (or whatever) had reached an infinite level.

Secondly, exponential curves are often drawn to look like the solid curve – usually to show that the growth rate under consideration has now reached fantastic proportions, and is going to become astronomical in the near future. But the shape of the curve can be misleading. The exponential curve possesses the fascinating property that if the vertical scale is expanded or contracted the new curve is the same shape as the previous curve at a later or earlier time. For example, if the solid curve were scaled down to one-quarter of its height it would look like the dotted curve. This is exactly the same shape as the solid curve earlier in its history, up to the point A. Thus by just expanding or contracting the scale of the axes one can readily shift the apparent point of crisis into the future or into the past. So the implications of an exponential curve should not be judged by its shape alone, but rather by the *actual* rates of growth as well.

The second aspect of diversity is increasing numbers. The human population has been rapidly expanding, and many see this as a negative trend. But from an evolutionary perspective, increasing numbers are vital, as they contribute to the complexity upon which evolution builds.

The growth of nearly all populations – whether they be

bacteria in a dish, cells in an embryo, or rabbits in Australia – initially follows a pattern of rapidly increasing growth typified by what is called the exponential curve. Since the nature of this curve is important for the discussions that follow, let us pause to take a look at some of its properties.

In exponential growth the rate of growth is directly proportional to the current size. This means that the bigger something is, the faster it grows. Population is a classic example. The rate of growth of a population depends upon the number of people already living, and so an expanding population tends to grow more and more rapidly (provided there are no sudden decreases caused by war, disease, famine, etc.). Another example of exponential growth occurs when money is invested at compound interest. The interest is calculated on the initial capital *plus* all the interest accrued so far; more and more interest is added each year, giving rise to the typical exponential curve shown in Figure 4.

Every exponential curve, because it grows at a constant percentage rate, has its own particular 'doubling time'. This is the time it takes for the population (or whatever is being measured) to double in size. A simple mathematical relationship exists between the percentage growth rate and the doubling time: doubling time equals 70 divided by growth rate (69.31 divided by growth rate to be a little more precise). Thus, if the population were growing at the rate of 2% per year, it would have a doubling time of 35 years. This relationship will be useful later when we come to consider the implications of the growth rates of various aspects of society.

The exponential curve is, of course, a mathematical model, and natural growths do not necessarily follow this pattern exactly. The growth of the human population, for example, has not followed a true exponential curve. Over the centuries the doubling time has steadily shortened. In the year 1000 A.D. the world population was around 340 million, with a doubling time of about 500 years. By the 17th century the doubling time had dropped to 300 years; in 1800 it was down to 100 years; in 1940 it was about 50 years; and in the 1960s it was doubling once every 35 years.

This type of growth – increasing more rapidly than even an exponential growth – is called 'super-exponential'. Such curves occur whenever the rate of growth not only depends on the current size of the population (which is the basis of exponential growth), but also builds on what has already been achieved in the past. (If, for example, a bank were to add up the sums that appeared on each of your annual statements, and pay you interest on that total rather than your current statement, you would have a super-exponentially increasing account, and soon be very rich indeed). In the case of the human population, this super-exponential growth is a direct consequence of the development of language, writing, printing and communication systems, which have brought with them the ability to pool the knowledge acquired over time. This has led to better health care, increased production, higher standards of living and more efficient land use – factors that have enabled the population to grow faster.

In practice, no exponential growth, nor any super-exponential growth, can go on increasing forever. Eventually it will reach limits imposed by the physical environment. Bacteria in a dish, doubling every few hours, cannot do more than fill the dish. Humanity cannot (at present) do more than fill the planet. At some point the growing population will start to feel the environment's inability to support ever-increasing

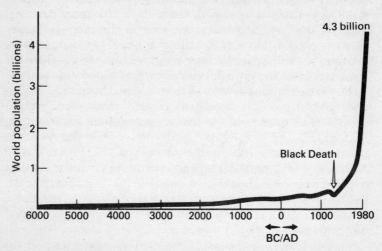

FIGURE 5. World population over the last 8,000 years.

numbers – usually when it is about halfway to the maximum. The growth rate will then begin to slow down, and the curve will start bending over in the opposite direction, producing an 'S'-shaped curve.

In the case of bacteria in a dish, this slowing of growth is brought about by such forces as lack of food and lack of space – factors that are beyond its control. In the case of the human population, we might expect similar forms of 'natural' control such as disease, famine, and perhaps even genocide resulting from conflicts over diminishing food, resources and energy supplies. Unlike the bacteria, however, humanity can anticipate the future and apply conscious decision-making to population control, giving us the possibility of pre-empting the various means of 'natural' control and avoiding apocalypse.

Recent data on the human population suggests that its growth has already begun to slow. The most accurate figures come from the developed nations, and in nearly all of these, fertility (defined as the average number of children born per woman) is steadily decreasing. Sweden has already reached zero population growth, while Germany (both West and East) and the USA are not far behind with fertilities below strict replacement level. (The fact that fertility may have dropped below replacement level does not necessarily mean that the population immediately stops growing. It will often continue increasing for as much as twenty years because the number of potential parents may still be growing as the result of earlier higher birth rates.) In China, with a quarter of the world's population, there is now a policy of one child per family. Reproduction there has likewise dropped below replacement level, and China expects to reach zero growth by the year 2000.

The data for the rest of the world is somewhat less reliable, but even so there seems to be a general trend towards steadily decreasing growth. Figures for the planet as a whole suggest that the annual growth rate reached a peak of 2% in the early sixties; by 1970 it had dropped to 1.9%, and by 1977 to 1.7%. We would now appear to be just past the mid-point of the 'S'.

Population analysts have made various predictions as to what will happen if these trends continue. The general concensus is that world population will reach around six billion by 2000 A.D., and will probably stabilise at somewhere between eight and eleven billion by the middle or end of the twenty-first century.

The possibility that the population may stabilize around 10^{10} is interesting. As we have already seen, this figure appears to represent the approximate number of elements that need to be gathered together before a new level of evolution can emerge. (There are about this number of atoms in a simple living cell, and this number of cells in the cortex of the human brain.) If the same pattern occurs at higher levels of integration, then the human race may be fast approaching the stage where there are sufficient numbers of self-reflective consciousnesses on the planet for the next level to emerge.

However, we may not have to wait for the end of the 21st century to see this possibility. After all, the number 10^{10} is not

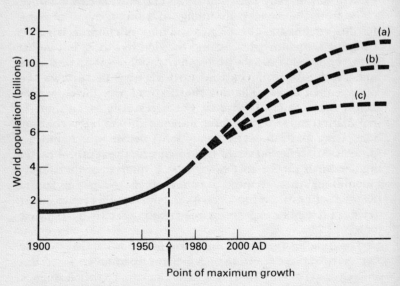

FIGURE 6. Predictions of future growth of world population by:
(a) United Nations,   (b) World Bank,   (c) University of Chicago,
showing slowing down of growth rate and eventual stabilisation between 8 and 12 billion.

(Note that the scales are different from those of Figure 5.)

an exact requirement; rather, it refers to a whole range of numbers that are of the same order of magnitude (i.e., which do not differ by more than a factor of ten). Thus the current population of four billion (4×10^9) is already well within the necessary range. So whether the population grows to eight or eleven billion is not important in evolutionary terms.

SOCIAL ORGANISATION

Numbers alone are not enough to bring about a major evolutionary leap. Ten billion assorted atoms put together on a pin point do not make a living cell, nor do ten billion neurons in a glass jar constitute a conscious brain. The elements need to be integrated into a cohesive structure and their interaction needs to be organized.

The first step towards this organisation is usually 'clumping' – the trend towards groupings of components. Considering society, we can trace a steady movement from small groups of nomadic hunter-gatherers, to farming communities; from tribal villages and hereditary clans, to small countries and states; and from nations to larger groups such as the USSR, the British Commonwealth and the EEC, which transcend geographic and racial boundaries.

As the groups have become larger and more integrated so have they become more organised. Just as a cell has its organelles, and the body its organs, so society has its organisations and structure. Rather than everyone doing much the same tasks, as in primitive hunter-gatherer societies, modern societies have led to a high degree of specialisation.

Today almost everybody is a specialist, and the resulting interdependence and interaction of human society has given rise to a highly complex social structure. Just to drive to the store for a bottle of orange juice depends upon an interconnected worldwide network of people working in such diverse places as a rubber farm, an oil well, a refinery, a steel mill, a copper mine, an automobile manufacturing plant, an orange farm, a glass company, and various import, export and distribution companies – to mention just a few.

The increasing organisation in society is not only found at

the physical level. With the evolution of human beings there emerged self-reflective consciousness and the ability to reflect upon the world we inhabit. This opened up the possibility of evolution at the mental level, and we can find the trend towards greater organisation manifesting within us in various ways.

Intelligence itself is an organising principle within human consciousness. In its most general sense, intelligence can be thought of as the ability to abstract order from raw sensory data, organise our perceptions into meaningful wholes, form relationships between them (concepts, expectations, hypotheses, etc.), and thereby organise action in a purposeful way.

The many facets of human knowledge can also be considered as ways of organising our experience of the world. Each individual scientific discipline represents a particular way of looking for underlying rules and laws; they are revealing the order of the world we live in. Art, likewise, attempts to bring to human consciousness hidden orders of creation. In these, and many other ways, human beings are continually discovering new relationships, increasingly organising their information about the world.

Human society has taken this one stage further. We not only organise information within ourselves, we can share that information with others. Through a variety of means of communication we are beginning to link together at the level of consciousness, thereby enhancing the third crucial aspect of complexity – connectivity.

THE INFORMATION AGE

To fully understand the significance of today's developments in the area of communications, we need to go back in time to see what social changes have occurred over the last 200 years. In this short period, the thrust of human activity has altered significantly.

Prior to the middle of the eighteenth century, the dominant area of human activity was the production of food (i.e., agriculture, fishing, and the systems for food distribution). The number of people employed in this field has generally increased at about the same rate as the population itself. In recent times, however, it has grown more slowly than the population due to the increasing application of technology to farming. As a result the growth curve for employment in food production has bent over into the characteristic 'S' curve.

Since the Industrial Revolution the more developed nations have shown a steady increase in the number of people employed in industry and manufacturing. This growth has been considerably faster than that of agricultural employment and an increasing proportion of the workforce became involved in industry. In general terms this shift could be considered as a shift from the processing of food to the processing of minerals and energy.

Whereas the number of people engaged in food production in the USA showed a doubling time of 45 years, industrial employment doubled about every 16 years. As a result the growth curve of industrial employment in the USA caught up with that of agricultural employment in the year 1900. In terms of employment, therefore, this date could be taken to mark the true beginning of the Industrial Age in the USA. From then on, industry became the dominant activity, and today, only 3% of the US population is engaged in food production.

In the last decade, the rate of growth of industrial employment has also slowed, again giving rise to an 'S' curve. This slowing has been the result of the steady application of technology and automation to industry.

Within the last twenty-five years or so, another major area of human activity has emerged – information processing. (This comprises fields such as education, printing, publishing, accounting, banking, journalism, TV, radio, telecommunications and the many types of activities based on computers.) Information processing is growing so fast that its doubling time may be as short as six years.

The rapidly expanding growth of communications has led to a further shift in employment – away from industry, the processing of energy and matter, towards the processing of information. By the mid-1970s the number of people in the USA engaged in this new field had caught up with those engaged in industry. Since that time, information processing

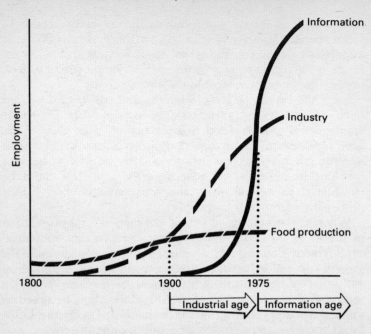

FIGURE 7. Changes in the number of people involved in different
categories of human activity: food production, industry and information
processing.

Employment statistics for the USA show that in 1800, prior to the
Industrial Revolution, 80 per cent of the work-force was employed in food
production. By 1900 the two had equalled, with about 38 per cent being
employed in both sectors. This shift has continued and today food
production takes up only 3 per cent of the employment.

Recently a new and more rapid growth has appeared – information
processing. It has a doubling time of around six years and in 1975 overtook
industry in terms of employment. Information is now the dominant form
of employment in the developed nations.

has become the dominant activity, marking the beginning of
the 'Information Age'.

Although these developments refer specifically to the USA,
parallel changes are to be found in most of the more developed
nations. The less developed nations show similar tendencies,
but they lag behind the more developed ones to varying

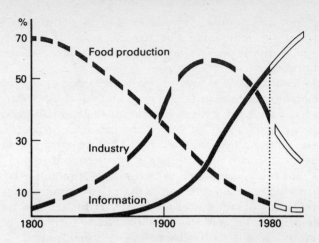

FIGURE 8. The same data as Figure 8, the number of people employed in agriculture, industry and information processing being plotted as the percentage of the total workforce. This brings out more clearly the relative changes in each area. In the late 1800s, about 50% were in agriculture; in the 1930s, over 50% were in industry; and in the 1980s, over 50% are in information processing. (Data applies to U.S.A.)

degrees. These lags, however, will almost certainly decrease as time goes on. While a country may be fifty years behind the West in reaching the stage at which industrial activity becomes dominant, it may only be ten years behind when it makes the transition to an information-dominant society. Japan is an example of a country that, despite its late start, has clearly caught up with the West. Many of the Middle East oil-rich nations, such as Kuwait and Saudi Arabia, are also making rapid strides in this direction. And China, although still predominantly agricultural, may spend only a short time in the 'Industrial Age' before shifting to an information society.

As more and more nations of the world move into the Information Age, the technology of communications and information processing will have a dramatic effect on the human race, as we become more and more integrated through the burgeoning network of electronic synapses.

If we look back over human history, we can see that this

trend towards a progressive linking of humanity has been going on for millennia. The sudden surge of information technology in the present day is the fruit of millions of years of human effort.

The first major step towards interconnection came with the development of human language. This lead to a more direct transfer of knowledge between individuals than had previously been possible, and facilitated the grouping of people into simple communities and villages. The second great breakthrough came with the invention of writing about 10,000 years ago. This made it possible to transmit information over time and space, encouraging the growth of large communities and the recording of cultural histories and traditions. The advent of the printing press in the fifteenth century further increased humanity's ability to disseminate written information.

The next major breakthrough occurred in the mid-nineteenth century. This was the development of electrical communication in the form of the telegraph, and later the telephone. People across the world from each other could now be linked together by electric cables, and the time taken to transmit a message over long distances suddenly dropped from days or weeks to fractions of a second.

Fifty years later another breakthrough occurred through the use of radio waves as the transmission medium. This freed people from the need to be physically linked by cable, and simultaneously made it possible to transmit a message to large numbers of people, i.e. to 'broad-cast' information. Since then, radio and its offshoot, television, have expanded rapidly, enabling the individual to be an eyewitness to events happening around the world.

At the same time as radio and television were spreading across the planet, another equally important development in information technology was occurring – computers.

Computers had their conception during World War II, when there arose the need to perform complex calculations and process information much more rapidly than could be done by paper, pencil and adding machine. To fulfill this need scientists designed electronic calculators, and these gave birth to early electronic computers in the 1950's. Although cumbersome and slow by today's standards, they nevertheless represented a huge leap forward in terms of information processing power and speed. During the 1960s and 1970s, dramatic strides were made in the computer's processing capacity and speed. Simultaneously the physical size of computers shrank remarkably.

The micro-processor, or 'chip' as it is commonly called, represented a major revolution in computing technology. Less than a half centimeter in size, the average chip of 1980 contains more computing power than all the computers of 1950 put together, and this capacity has been doubling every year. In addition to the many advantages of its minute size, the chip's energy consumption is astoundingly low. The average computer of 1970 used more energy than 5,000 pocket calculators of similar computing capacity a mere ten years later. As a consequence, the information/energy ratio has been steadily increasing, and is now rocketing. We are able to do more and more with less and less.

Whereas in 1970 computers were used almost solely by large institutions such as governments and corporations, the chip has made it possible for the technology of computers and data processing to become potentially available to anyone on the planet, and to do this without draining the planet of its vital energy resources. If comparable changes had been made in various aspects of the automobile over the last 20 years, then a Rolls Royce would now cost two dollars, it would be less than an inch long, have a petrol consumption of several million miles per gallon, cruise at a hundred thousand miles per hour, and never need servicing!

Another significant development has been the direct linking of computers. The first computers were independent units, interacting only with human beings. In the late 1960s, however, two computers were linked together, making it possible for them to communicate directly. Within only a few years this development had given birth to numerous networks of computers exchanging data across the world, very much more rapidly than humans ever could. By 1980 thousands of such networks were in operation, being used for everything from stock control and airline reservations to crime prevention and scientific research. Moreover, the number of networks is doubling every two to three years.

Although computer usage was initially confined to mathematical calculations and data processing, chips and networks have allowed computers to be applied to the field of human communications. What before was transmitted through the media of print, voice, records, tapes, or pictures can now be translated into the language of computers (i.e., a series of 'o's and '1's called 'bits'), transmitted as electronic data, then 'reconstituted' at the receiving end, and with near perfect fidelity.

As the marriage between communication technology and computers develops further, electronic mail, newspapers and shopping will become commonplace, as will videophones with multiway conversations. Through home computer terminals, no larger than a pocket calculator, each person will be able to be in contact with numerous other people, with other computers, and with data banks across the world. With the further application of simultaneous data transmission and processing, many more new horizons will open up, some of which we may not be able even to imagine.

Progress in computer technology is described in terms of 'generations'. Each new generation is based on fundamentally new technological advances. The first three generations of computers were built from vacuum tubes, transistors and integrated circuits respectively. Chips are fourth generation machines. The fifth generation, using a special phenemenon called the Josephson Effect (after its discoverer Nobel laureate Brian Josephson), will probably appear in the mid-1980s. Operating at temperatures near absolute zero, these computers will be about a hundred times faster than current machines. When the sixth generation appears, probably in the 1990s, we may well have computers that could equal a human brain in any intellectual activity.

It would then only be a small step to what is termed the UIM (ultra-intelligent machine), a machine that could do as well or better than a human being in any field of information processing. Once this stage is reached, we will see an even more fantastic acceleration in computing power. UIMs could be put to the task of designing better computers, and second generation UIMs will be built with processing capabilities way beyond human intellectual abilities. Third and fourth generation UIMs will doubtless follow, with exponentially increasing intelligence.

As we contemplate future scenarios, one that quickly comes to mind – a favorite of a number of science fiction writers – is that computers would eventually become so intelligent that they would take over from humanity as the spearhead of evolution, and humanity would then become obsolete. Such a development, though it would represent the further evolution of intelligence, would not however be in line with the overall pattern of evolution.

In the past each new development has come from the integration and improvement of existing systems. At no stage has any branch of evolution been replaced by its by-products or artifacts; they have always evolved together, becoming more integrated rather than separate. If the pattern of the past 4.5 billion years is followed, the next stage will involve the co-evolution of humanity and its artifacts in a mutually supportive relationship. Rather than UIMs taking over from humanity, they will become another factor increasing the integration of society and speeding human evolution towards greater complexity.

THE EVOLUTION OF A GLOBAL BRAIN

The embryonic human brain passes through two major phases of development. The first is a massive population explosion of the embryonic nerve cells, starting eight weeks after conception. During this phase, the number of cells increases by many millions each day. After five weeks, however, the process slows down, almost as rapidly as it started. The first stage of brain development – the proliferation of cells – is now complete.

From there the brain proceeds to the second phase of its development, as billions of isolated nerve cells begin making connections with each other, sometimes with neighbouring cells, sometimes growing out fibres to connect with cells on the other side of the brain. By the time of birth, a typical nerve cell may communicate directly with several thousand other cells, and some cells with as many as a quarter of a million. This proliferation of connections continues through the first years of life.

Similar trends can be observed in human society today. We seem to be moving out of the period of massive 'cell' proliferation and into a phase of growing interconnectivity. As worldwide communication capabilities become increasingly complex, society is beginning to look more and more like a planetary nervous system. The global brain is being activated.

This awakening is not only apparent to us, it can even be detected millions of miles out in space. Before 1900, any being curious enough to take a 'planetary EEG' (i.e. to measure the electromagnetic activity of the planet) would have observed only random naturally-occurring activity, such as that produced by lightning. Today, however, the space around the planet is teeming with millions of different signals – some of them broadcasts to large numbers of people, some of them personal communications, and some of them the chatter of computers exchanging information. As the usable radio bands fill up, we find new ways of cramming information into them, and new spectra of energy, such as light, are being utilised, with the potential of further expanding our communication capacities.

With near-instant linkage of everyone to everyone through this communications technology, and the rapid and wholesale dissemination of information, Marshal McLuhan's vision of the world as a 'global village' is fast becoming a reality. From an isolated cottage in a forest in England I can dial a number in Fiji, and it takes the same amount of time for my voice to reach Fiji down the telephone line as it does for my brain to tell my finger to touch the dial. As far as time to communicate is concerned, the planet has shrunk so much that the other 'cells' of the global brain are no further away from our brains than are our own bodies.

At the same time as the speed of global interaction is increasing, so is the complexity. In 1980 the worldwide telecommunications network consisted of 440 million telephones, and nearly one million telex machines. Yet this network, intricate as it might seem, represents only a minute fraction of the communication terminals in the brain, the trillions of synapses through which nerve cells interact.

According to John McNulty, a British computer consultant, the global telecommunications network of 1975 was no more complex than a region of the brain the size of a pea. But overall data-processing capacity is doubling every two and a half years, and if this rate of increase is sustained the global telecommunications network could equal the brain in complexity by the year 2000 – if this seems an unbelievably short a time ahead, it is probably because few of us can really grasp just how fast the growth is.

The changes that this will bring will be so great that their full impact may well be beyond our imagination. No longer will we perceive ourselves as isolated individuals; we will know ourselves to be part of a rapidly integrating global network, the nerve cells of an awakening global brain.

CHAPTER 6

The emergence of a social super-organism

We no more know our own destiny than a tea leaf knows the destiny of the East India Company.

Douglas Adams, *The Hitchhiker's Guide to the Galaxy*

The growing complexity that we have just traced within society reveals three important areas of growth in terms of evolution: a diversity of human beings, their numbers stabilizing around the critical size of 10^{10}; an elaborate organizational structure, parallel to that observed in all other living systems; and a communication and information processing capability approaching that of the human brain. Thus society would appear to be completing the prerequisites for the emergence of a new evolutionary level.

What might this new level look like?

Just as matter became organised into living cells, and living cells collected into multi-cellular organisms, so might we expect that at some stage human beings will become integrated into some form of global social super-organism. (I use the word super-organism rather than organism since the term 'organism' strictly applies to biological organisms, i.e. to multi-cellular organisms, whereas we are now talking of multi-organism organisms, and these may take very different forms from biological organisms.)

A social super-organism, in the sense used here, is more

than just a living system. We have already seen in Chapter One that human societies display each of the nineteen characteristics of a living system. The same applies to many other social groups; a ship's crew, a factory and a multinational company, for example, may also be considered as living systems. But they are not social super-organisms in the sense of being independent wholes.

If we compare a social super-organism to a biological organism such as the human body, these social groups are more like constituent organs such as the thyroid gland, an eye and the liver. These organs are living systems, but they can only exist as part of a larger organism, within which they play specific roles. On its own a thyroid gland, eye or liver cannot survive. Nor can a ship's crew, a factory or a multinational company survive alone. They are part of a larger system, within which they have specific functions. A social super-organism, on the other hand, would, like a biological organism, be an independent whole, complete in itself.

Large countries might seem to be more like super-organisms in that they can function as independent units. But, as we shall see shortly, one further feature, crucial to any healthily functioning organism is severely lacking in nearly all human societies.

Social super-organism are not new to nature. In the animal world there are several examples of organisms that come together to form integrated social units. Thousands of bees may live and work together in a single hive, regulating the temperature and humidity of their collective 'body', the colony as a whole surviving the continual birth and death of its members. Army ants form colonies containing up to 20 million individuals. Advancing like a single organism through the forest, a colony will cross streams by forming a living bridge of ants clinging tenaciously to each other. When one army meets another they will avoid close contact, behaving like two separate entities. Termites construct complex cities housing up to several million individuals, complete with ventilation shafts and complex food-processing systems.

Similar tendencies can be found in higher animals. Many fish swim in schools, the whole acting as a single unit without any single leader. Individual fish may take on specific functions, such as that of an 'eye', reducing the need for the others to be continually on the lookout. When danger is spotted a whole school can react in less than a fifth of a second. Flocks of birds can likewise behave like a super-organism. (One of the largest ever recorded was a flock of 150 million shearwaters, over ten miles across, observed between Australia and Tasmania.) Slow-motion films of bird flocks have revealed 50,000 individuals turning in synchrony in less than a seventieth of a second. There is no indication of their following the leader; the flock is integrated into a functional whole.

But such examples, fascinating as they might be, only afford us a glimpse of the integrated social super-organism that humanity has the potential to become. Firstly, this super-organism will not contain a few million individuals, as occurs with bees, ants, or birds; rather, it will be comprised of the whole human race – billions of individuals distributed over the face of the planet.

Secondly, in all instances of animal super-organisms there is very little individual diversity. Bee and ant colonies usually contain only two or three different types (e.g., worker bee, drone, queen bee), while in fish and bird groups all the individuals are usually identical, only temporarily taking on specific functions such as an 'eye' or 'skin'. Human society, however, is extremely diverse and specialized, made up of thousands of different types, each able to make his own particular contribution to the whole.

Thirdly, a human social super-organism would not entail our all becoming non-descript 'cells' who have given up their individuality for some higher good. We already are 'cells' in the various organs that compose society, yet still retain considerable individuality. The shift to a social super-organism would essentially mean that society became a more integrated living system. As we shall see in later chapters, this is likely to lead to greater freedom and self-expression on the part of the individual, and to an even greater diversity.

Lastly, when insects and animals come together, they congregate as a single unit. But it is extremely unlikely that the human social super-organism will form itself on the physical level. From what we have seen of evolutionary trends, we should not anticipate that human beings will come together as a large conglomerate mass in some supermegalopolis. With humanity evolution has gained a new platform on which to build.

Just as earlier, after life had emerged from matter, evolution moved up from the physical level to the biological level, so it now has moved up to a new level – consciousness. We could therefore hypothesize that the integration of society into a super-organism will occur through the evolution of consciousness, rather than through physical or biological evolution. This implies a coming together of minds, which is why communication is such an important aspect of evolution today; it is a mind-linking process. Humanity is growing together mentally – however distant we might be physically.

THE FIFTH LEVEL OF EVOLUTION

One philosopher who spent much of his life contemplating the integration of humanity into a single being was Pierre Teilhard de Chardin. Teilhard displayed that rare synthesis of science and religion: he was both a Jesuit priest and a geologist/paleontologist. In the 1930s he worked in China where he was closely involved with the discovery of the skull of 'Peking Man'. As a result of his study of the evolutionary process, he developed a general theory of evolution, which he believed would apply not only to the human species but also to the human mind, and the relationship of religious experience to the facts of natural science.

One of his principal conclusions was that humanity was headed towards the unification of the entire species into a single interthinking group. He coined the word 'noosphere' (from the Greek *noos*, 'mind') to refer to the cumulative effect of human minds over the entire planet. Just as the biosphere denotes the system comprising all living beings, so the noosphere denotes that comprising all consciousness minds.

Evolution, having passed through geogenesis (the genesis of the Earth), and biogenesis (the genesis of life), was now at the stage of 'noogenesis' (the genesis of mind). He saw this stage as the 'planetization of Mankind . . . (into) a single, major organic unity'. The fulfilment of the process of noogenesis Teilhard referred to as the 'Omega Point', the culmination of the evolutionary process, the end point towards which we are all converging.

Another philosopher with a similar vision was the Indian mystic Sri Aurobindo, a contemporary of Teilhard's. We also find an interesting combination of talents in Aurobindo. He was educated in classics at King's College, Cambridge, and was, on his return to India, an active political revolutionary. As a result he spent several years in Indian jails, and it was during these periods that he had some of his most significant insights into the evolution of humanity.

Sri Aurobindo saw evolution as the 'Divine Reality' expressing itself in ever higher forms of existence. Having passed from energy through matter and life to consciousness, evolution was now passing through the transformation from consciousness to what he called 'Supermind', something so far above consciousness as to be beyond our present dreams of perfection – the ultimate evolution of 'Spirit'. This new level he saw as coming through the increasing spiritual development of individual consciousness towards a final, complete, all-embracing consciousness, which would occur on both the individual and collective levels.

When Teilhard wrote of the unification of the noosphere in the Omega Point, he appears to have been talking about an integrated planetary consciousness. Similarly, Aurobindo, although he made it clear that Supermind was far above the individual mind, still generally discussed this new phenomenon in terms of mind and consciousness.

Yet the evolutionary trends and patterns we have lookd at thus far suggest a further possibility: the emergence of something way beyond even a single planetary consciousness or Supermind – a completely new level of evolution, as different from consciousness as consciousness is from life, and life is from matter.

This new order of existence will be the ultimate effect of the continued integration of humanity, and in this respect humanity will be its platform. Yet it will not be happening to humanity. Rather, it will be happening at the planetary level, to Gaia. (A helpful analogy might be made with a photograph composed of many dots. The dots themselves do not contain the picture, yet they are its basis. The picture emerges from the collected organisation of the dots viewed as a whole.)

Since we do not yet have an adequate term in our vocabulary for this fifth level, I have chosen to refer to it as the 'Gaiafield' (much as self-reflective consciousness might be termed the 'human field'). The Gaiafield will not be the property of individual human beings, any more than consciousness is the property of individual cells. The Gaiafield will occur at the *planetary* level, emerging from the combined interactions of all the minds within the social super-organism.

Exactly what the nature of this new level would be is very difficult to say. The minute we begin to contemplate a fifth level of evolution, we inevitably do so in terms of human experience. But as we have already seen, each new order of existence is not fully describable in terms of the previous orders, and the Gaiafield would possess entirely new characteristics incomprehensible to consciousness.

Similarily, a single cell in your body knows nothing of the consciousness that emerges from the living system as a whole. Although it might have a very rudimentary form of awareness, it has no conception (if we may excuse the word) of your thoughts, feelings, imaginations and inspirations. It does not know what state of consciousness you are in, or even whether you are conscious or not. And it would almost certainly find it impossible to 'conceive' of what was meant by self-reflective consciousness.

Thus it is not altogether surprising if we find it equally difficult to conceive of evolutionary stages as far beyond us as we are beyond single cells. Being individuals, the collective phenomena must theoretically remain to us an unknowable. We can only know the 'cell' that we are.

The idea of a collective phenomenon generated from the activity of our individual consciousnesses may sound a little strange. How can many separate consciousnesses give rise to a single planetary phenomenon? The question is no easier to answer than the one which scientists and philosophers continually face with respect to human consciousness: how do the electrical and chemical activities of many separate nerve cells give rise to a single integrated consciousness?

All that can be said with any degree of certainty is that the consciousness of each individual human being is somehow related to the highly complex and integrated interaction of billions of living cells in the brain. Without going into the neurological arguments here, it seems probable that a particular conscious experience corresponds not to the activity of small groups of cells, but to overall patterns of activity, to the coherence of the numerous information exchanges continually taking place in the brain. In a similar manner this planetary field would emerge from the integrated interaction of the billions of conscious beings composing humanity. As the communication links within humanity increase, we will eventually reach a time when the billions of information exchanges shuttling through the networks at any one time create similar patterns of coherence in the global brain as are found in the human brain. Gaia would then awaken and become her equivalent of conscious.

How far away are we from the point in time where this might happen? Teilhard, although he spoke of evolution moving rapidly towards the Omega Point, was thinking in cosmic timescales rather than human ones, and by 'rapidly' he seems to have meant thousands or perhaps millions of years. Sri Aurobindo, on the other hand, believed that it could happen a lot more quickly, perhaps within the next hundred years.

Yet this new level may well come even faster than either Teilhard or Sri Aurobindo envisaged. It could possibly happen within a few decades.

We have been accustomed to social changes occurring over centuries, or even millennia, and it may be difficult to conceive of something so significant emerging in so short a period of time. But, as we saw in Chapter Four, the rate of change is increasing rapidly. In this century there has been a massive acceleration in nearly all areas of human endeavor, heralding some form of major transition in the very near future. Furthermore, the 'major evolutionary indicators' (diversity, organization, and connectivity) are rapidly producing the critical degree of complexity that seems to be needed for a new level to emerge. We shall also find, in Part Two, that there are several other reasons for supposing that we could well see this shift happening in our own lifetimes. We may well be about to experience a break with historical patterns.

Such a transition is clearly going to require some very rapid changes on the part of humanity. One need only pick up a newspaper to see just how far humanity is from being a cohesive, integrated whole. Yet society is much more than a collection of disassociated individuals going their separate ways. We find ourselves in a sort of 'twilight zone', neither one thing nor the other. As it turns out, this is a characteristic phase of evolutionary transitions.

EVOLUTION'S TWILIGHT ZONES

The boundaries between the major evolutionary levels (i.e. energy, matter, life and consciousness) are not necessarily clearcut. In between any two appear to be twilight zones, where the new order is becoming manifest but has not yet fully emerged.

Looking first at the junction between energy and matter, we find the so-called elementary particles, such as electrons and protons. But are they really matter? In some situations they behave like particles, yet in other situations they behave like waves – behaviour more characteristic of energy – a paradox that many physicists have tried to resolve. But when we consider evolution as an emergent process, an elementary particle can be seen to stand on the borderline between energy and matter; it is matter *emerging* from energy, a halfway stage.

Between the next two levels, between matter and life, are the viruses and macromolecules such as DNA. Here the question is: Are they alive? In some respects they are, since they can, under suitable circumstances, reproduce themselves. But they can also form regular crystals, behaving like many of the simpler molecules. They seem to be in the twilight zone where matter has not yet become sufficiently organised to lead to the emergence of true life.

A similar twilight zone exists between life and self-reflective consciousness. In this case it is occupied by primates such as the chimpanzees. Like other animals, they are conscious, in the sense of being aware of their surroundings. But are they conscious of themselves, and conscious that they are conscious? The answer is not clearcut. In experiments where chimps have been taught a simple sign language, some appeared to display a rudimentary self-consciousness in being able to refer to themselves by name (although the evidence for this is still very debatable). Without such a language, however, chimps – intelligent and expressive as they are – show little indication of being conscious that they are conscious. Nevertheless, chimpanzees, and a few other primates such as orang-outangs, are different from all other animals in that they display self-recognition. When they see an image of themselves in a mirror they realise they are seeing themselves rather than another animal of the same species. They would appear, therefore, to be somewhere in the twilight zone that preceeds the emergence of true self-reflective consciousness.

At the present time, evolution appears to have reached yet another twilight zone – the one between consciousness and the Gaiafield. Humanity currently displays characteristics of both levels: we are independent conscious units, and at times we come together to function as an integrated whole for a common purpose. In this respect, society is reminiscent of a curious creature, the cellular slime mould (*Dictyostelium discoideum*), an 'organism' which is somewhere between a collection of single-celled amoebae and a true multi-cellular organism.

Most of the time the separate amoebae constituting a slime mould roam around old bits of wood and dead leaves, looking for bacteria on which to feed, and multiplying as they go. Should the food supply become scarce, the separate amoebae start clustering into small groups of a few dozen individuals. These groupings then conglomerate into a single blob, called a 'grex', often containing thousands of amoebae. Having come together, some of the cells begin climbing on top of the others until they form a hemispherical dome, which develops into a cone with a 'nipple' on top. The whole thing then falls over onto its side and becomes a small 'slug', able to move across the forest floor in the direction of light, the nipple raised and leading the way.

If food is found, the grex may dissolve again into thousands of individual amoebae going their separate ways. If not, it may turn up on its end, thousands of amoebae climbing on top of each other to become a thin vertical stalk as much as an inch in height. At the top of this thin stalk other amoebae will form into a small sphere and become spores to be cast off and carried away in the air. If an amoeba lands somewhere with an abundant supply of food, it begins reproducing and spreading as before – until, that is, the food supply begins to become exhausted, whereupon the whole process begins over again.

Parallel behaviours are found in human societies, both primitive and advanced. The Kachin people of north Burma, for example, who have been studied extensively by the British anthropologist Edmund Leach, spend most of their time in separate tribal communities. When food is scarce, however, they come together as a unit under one king and stay as a single community until times improve. Similarly in the more developed Western nations, when there are no major catastrophes, each person mainly pursues his own interests. But should there be a disaster, such as a widespread famine, flood or war, people begin acting more in the group interest, and society takes on more of the characteristics of an integrated organism.

EMERGENCE THROUGH EMERGENCY

To say that humanity is in the twilight zone is not to imply that the emergence of the next level is inevitable. Transition periods are fraught with danger, and this is clearly the case with society today. We are deeply entangled in the most complex web of social, political, economic, ecological and moral crises in human history. Will these crises prevent the emergence of a new level of evolution? Perhaps. Certainly we have any number of doomsday projections which explore the possibilities of apocalypse in detail. But our earlier review of evolution revealed quite another possible scenario – that crisis may be an evolutionary catalyst in the push towards a higher level.

Initially, any crisis looks painful and dangerous, and one's immediate reaction may be to try to stop it, holding on as firmly as possible to the old order. But if there is the possibility of a new order emerging from this crisis, hanging on to the current state may be counterproductive, perhaps even deepening the problem. Imagine what a committee of bacteria would have said about the environmental impact of a small group of bacteria's plans to use photosynthesis, 3,500 million years ago: 'The oxygen produced by this process is dangerous stuff. It is poisonous to all known forms of life and it is also highly inflammable, likely to burn us all to ashes. It is almost certain to lead to the destruction of life.' Without doubt photosynthesis would have been banned as 'selfish, unnatural and irresponsible'. Luckily for us, no such committee existed, and photosynthesis went ahead. It did indeed bring about a major crisis, but on the other side of it came plants, animals, you and me.

The set of global problems which humanity is now facing may turn out to be of equal importance to our continued evolution as was the oxygen crisis. Never in the history of the human race have the dangers been so extreme; yet in their role as evolutionary catalysts, they may be just what is needed to push us to a higher level.

The idea that crises have both negative and positive aspects is captured in a word the Chinese have for crisis: *wei-chi*. The first part of the word means 'beware, danger'. The second part, however, has a very different implication. It means 'opportunity for change'.

The concept of *wei-chi* allows us to appreciate the importance of both aspects of crisis. In recent years, our attention has generally been focused on the '*wei*', on the many possibilities for global catastrophe and how to avoid them. This will continue to be necessary as we strive to deal with the very real problems that face us. At the same time, these crises may lead us to question some of our basic attitudes and values: Why are we here? What do we *really* want? Isn't there more to life? This questioning opens us up to the other aspect of crisis, *chi* – the opportunity to change direction, to benefit from the prodigious and breath-taking possibilities that could be before us.

The journey so far

We started our exploration with the view of our planet seen from space and asked whether it could be a living system in its own right. If so, what is humanity's role within it? Does our ability to collect, store and disseminate information make us somewhat like the nervous systems of the planet? Or are we more like a planetary cancer, blindly destroying the very environment on which we are so dependent?

Surveying the history of evolution from the Big Bang through to the emergence of modern civilisation, we saw a general trend towards greater complexity – increasing diversity, organisation and connectivity. When certain levels of complexity are reached, new levels of evolution emerge, with qualities entirely different from what has come before. Thus physical matter emerged from energy, life emerged from matter, and self-reflective consciousness emerged from life.

Evolution also shows a tendency to accelerate, particularly before the emergence of a new evolutionary level. Today the rate of change in society is accelerating so fast that it is difficult to predict the future even a decade or two ahead. Moreover, many of humanity's achievements represent evolutionary developments as significant as the earlier invention of photosynthesis or sexual reproduction. We are living through the most dramatic and crucial period of human history so far. We appear to be on the threshold of a major evolutionary transition.

If the patterns of the past continue, we should expect the

next major step in evolution to be integration of humanity into a single whole – a social super-organism. This super-organism might well be like some form of global brain. Already there are a similar number of people on the planet as there are cells in the human brain (that magic number 10^{10}). Furthermore, the rapid development of the global communications network suggests that humanity is fast approaching a similar degree of connectivity as that found in the brain. When this occurs we may see the emergence of a new level of evolution – the Gaiafield – the planet's equivalent of consciousness.

`Whether we do make such a transition is, however, far from certain. Humanity is clearly in a time of severe crisis and cannot continue on its present path for very much longer. Yet from an evolutionary perspective crises can be the impetus to new levels of order and organisation – if, that is, the system can make the necessary adaptions to survive the crises. Clearly more is needed than humanity linking up to form an integrated global nervous system. It is apparent that despite this increased communication, the problems are far from disappearing.

Ten billion neurons, even with their trillions of connections, do not necessarily result in consciousness – something more is needed. Similarily, even though the several billion people on the planet are connecting into a global nervous system something more is needed before an integrated social super-organism can emerge. It is to this missing ingredient, and to the ways in which it can be developed, that we shall now turn.

PART TWO

Inner Evolution

People travel to wonder at the height
of mountains, at the huge waves of the
sea, at the long courses of rivers, at
the vast compass of the ocean, at the
circular motion of the stars; and they
pass by themselves without wondering.

St.Augustine. 399 A.D.

Synergy

What we have traced so far is society's increasing complexity, and the many indications that we could now be living through the most significant, dramatic and crucial period of human history – the progressive integration of human minds into a single living system.

Yet we do not have to look far to see that humanity today is also on the brink of disaster. Paradoxically, the very same technological, scientific and social advances that have pushed us so far forward may also contain the seeds of our demise. We appear to be at a historical cusp, wavering between two mutually-exclusive directions: breaking through to become a global social super-organism; or breaking down into chaos and possible extinction.

Clearly, if given the choice, most people would not consciously opt for catastrophe. Nevertheless, as a group we seem to be drifting in that direction. Unable to fathom the complexity of the society we have become, we seem powerless to steer it in the direction we would want it to go. Why is this? Why are we not more like the organism we have the potential to be?

The answer lies in what characterizes a successfully-functioning organism. When we look at organisms that work – and just about every organism apart from human society does work – we find that there is one particular quality which they all share: the many components naturally and spontaneously function together, in harmony with the whole. This characteristic can be seen operating in organisms as different as a slime mould, an oak tree, or the human body. It is usually described by the word *synergy*, derived from the Greek *syn-ergos*, meaning 'to work together.'

Synergy does not imply any coercion or restraint, nor is it brought about by deliberate effort. Each individual element of the system works towards its own goals, and the goals themselves may be quite varied. Yet they function in ways that are *spontaneously* mutually supportive. Consequently, there is little, if any, intrinsic conflict.

The word 'synergy' has sometimes been used in the sense of the whole being greater than the sum of its parts. But this is not the word's root meaning; this interpretation is a *consequence* of synergy in its original sense. Because the elements in a synergetic system support each other, they also support the functioning of the system as a whole, and the performance of the whole is improved.

An excellent example of a system with high synergy is your own body. You are an assortment of several trillion individual cells, each acting for its own interest, yet each simultaneously supporting the good of the whole. A skin cell in your finger is doing its job as a skin cell, taking in various nourishments, getting rid of its waste products, and living and dying as a skin cell. It is not directly concerned with what is happening to a skin cell in your toe, nor to what is happening to your bone cells, blood cells, brain cells or muscle cells. It is simply looking after its own interests. Yet, its own interests are also the general interests of other cells in the body, and the activity of the organism as a whole. If it were not for this high synergy, each of us would be just a mass of jelly – each cell acting only for itself and not contributing to the rest of the body.

Synergy in an organism is the essence of life, and it is intimately related to health. When for some reason synergy drops and the organism as a whole does not receive the full support of its many parts, it becomes ill. When synergy is lost altogether, the organism dies. The individual cells may live on, but the whole – the *living* organism – no longer exists.

Likewise in social groups, synergy represents the extent to which the activities of the individual support the group as a whole. Anthropologists studying primitive tribal systems have found that groups high in synergy tend to be low in conflict and aggression, both between individuals and between the individual and the group. This does not mean that such societies are full of 'do-gooders' desperately trying to help each other; they are societies in which the social and psychological structures are such that the activity of the individual is naturally in tune with the needs of others and the needs of the group.

Viewed as a system, human society today would appear to be in a state of comparatively low synergy. And, as we shall see shortly, many of the crises now facing us are symptomatic of this deeper, underlying problem. Yet as much as we might want increased synergy in society, it is not going to come about simply through desire, intellectual decision, argument or coercion. The amount of synergy in society is a reflection of the way in which we perceive ourselves in relation to the world around. In order to increase synergy, we will need to change some fundamental assumptions which lie at the core of our thinking and behaviour. This will mean evolving inwardly as much as we have done outwardly.

The spearhead of evolution is now self-reflective consciousness. If evolution is indeed to push on to yet higher levels of integration, the most crucial changes are going to take place in the realm of human consciousness, and consciousness of the self in particular. In effect the evolutionary process has now become internalised within each of us. To see what this means, and how we may evolve inwardly, let us start by looking at how our internal model of ourselves governs our perception, thinking and action.

CHAPTER 7

The skin-encapsulated ego

Two birds
   inseparable companions
   perch on the same tree.
One eats the fruit,
   the other looks on.

The first bird is our individual self,
   feeding on the pleasures and pains of this world;
The other is the universal Self,
   silently witnessing all.

Mundaka Upanishad

For thousands of years people believed that the sun went around the Earth. So widespread and firm was this belief that it was taken to be 'reality.' In the sixteenth century, however, Copernicus put forward the radically different idea that the Earth went around the sun. But his theory was not readily accepted. It took a century of haggling before the old 'reality' was scrapped and a new 'reality' adopted.

This complete reversal in worldview did not come about through the discovery of any fundamentally new data, but through the interpretation of existing data in a new way. Clearly, the motion of the planets had not changed; what changed was the conceptual perspective through which their motion was viewed.

Scientists speak of this process as the creation of a new paradigm, a term coined by the philosopher and science historian, Thomas Kuhn in his book *The Structure of Scientific Revolutions*. The word 'paradigm' (coming from the Greek *paradigma*, meaning 'pattern') was used by Kuhn to refer to the dominant theoretical framework or set of assumptions which underlies any particular science. A paradigm is like a 'super theory'. It provides a basic model of reality within a particular science. It also governs the way a scientist thinks and theorises, and the way in which experimental observations are interpreted.

Once accepted, paradigms are seldom questioned; they usually become self-perpetuating scientific dogma. As a result, scientists will tend to accept those phenomena that fit in with the model and reject those that do not. However, there occasionally comes a time when the phenomena that do not fit become so well established that they can no longer be ignored. This usually results in what Kuhn referred to as a paradigm shift – something which we will be looking at in more detail later.

Although Kuhn originally put forward the idea of paradigms in relation to scientific thought, his ideas have been applied to many other areas: education, economics, sociology, politics, health care and our world view in general. The principles can also be applied to the way in which we perceive 'reality' and relate to ourselves.

Underlying our thoughts, perceptions and experiences are implicit assumptions about the way the world is. In seeing, for example, the eyes supply the brain with sensory data about the world 'out there'. But before this data can give rise to a meaningful experience, it first has to be interpreted and organised by the brain, and this requires a model of the world (i.e. an idea of how things are). On its own, without these perceptual frameworks, the raw visual data remains meaningless, as is shown by the following illustration.

To most people Figure 9 will appear as a rather random assortment of black and white patches. But it can be seen as a face, a rather medieval-looking bearded man. Without a visual model of what the face looks like, however, few people are able to see it at first glance.

FIGURE 9. A random array of black and white patches? An aerial photograph of Baffin Island? It can be seen as a face, but without a mental model of how the face looks, it can be difficult to see it.

Look at the illustration for a minute or so and if you still cannot see the face (as most people cannot) then take a good look at the illustration on page 116. You should now have a satisfactory visual model with which to interpret the incoming sensory data, and so 'see' the face. Moreover, you will probably always see it so long as you remember what it should look like, that is, so long as you retain the model.

Psychologists refer to the mental models that underlie the construction of perception as 'sets'. They not only condition most of our experience, they also determine what is 'reality' for each of us. We are predisposed by our sets to see certain features in our environment more than others. If, for example, you have just bought a new car, you will probably start seeing a lot more of those cars on the streets, particularly ones of the same colour. You might think the market had suddenly been flooded with them. Yet the number of such cars has not changed; what has happened is that your mind has become 'set' for them and notices them much more readily.

Mental sets can influence our behaviour and performance. An athlete who is convinced he can set a new world record, for example, is more likely to achieve this goal than an athlete of equal capability who has the set that the record is virtually unbeatable.

They can also affect our emotional reality. A depressed person, who feels that no one respects or loves him, and that the whole world is out to get him, has a negative mental set. Experiences and conversations are interpreted in a pessimistic light; positive, supportive comments are dismissed or undervalued; and the personal reality of gloom becomes self-reinforcing.

Similarly, the way in which we evaluate the world is affected by set. If our general set is of impending economic collapse, of international tension and aggression leading sooner or later to World War III, of potential disasters and famines, then we will be more likely to notice those elements when reported by the media, with the effect that the negative set will be reinforced. Moreover, we are also likely to act in ways that support that set. Like the athlete who does not really believe he can set a new record, our efforts to change the world will be half-hearted. We will create a self-fulfilling prophecy.

In short, mental sets, whether we are aware of their presence or not, are extremely powerful. They determine how sensory data is to be interpreted, which experiences to accept as 'real' and which to reject as 'illusion', and what 'reality' is like. Like paradigms, they are usually taken for granted, and seldom, if ever, questioned.

SELF-MODELS

Underlying how we think the world works (paradigms) and how we construct our experience (mind sets) is an even more basic model: the way in which we see our selves and the relationship between this self and everything else. This fundamental model conditions *all* thought, perception and action. It is the mind set, or paradigm, for all mental activity. Furthermore, since a self-model is often implicit in many educational, social, economic and political paradigms, it can even condition the development of paradigms themselves. If a physicist, for instance, experiences his consciousness and the physical world to be completely separate entities, he is likely to evolve different paradigms than he would if he experienced the two as part of a greater whole. In this respect our self-model is far more than a set or paradigm. It should strictly be termed a *meta*set (from the Greek *meta*, 'beyond') or *meta*paradigm – that which lies beyond all other sets and paradigms.

The most common self-model – the one by which most of us operate – is that of an individual self quite separate and distinct from the rest of the world. Functioning within this model, we go about our daily lives with the assumption that 'I' am 'in here', while the rest of the world is 'out there'. The philosopher and theologian Alan Watts dubbed this the 'skin-encapsulated ego': what is inside the skin is 'me', what is outside is 'not me'. All our perceptions and experiences are interpreted on this basis, and we model reality accordingly. This view of the self is so pervasive that few people ever realize it is just a model, or notice its effects on their experience and thinking.

But this model of self is not the only one. A radically different yet complementary model is possible – that of a universal self, not bounded by the skin, a self whose essential quality is oneness with the rest of creation, rather than separation from it.

Although momentary experiences of the universal self are not as rare as one might suppose, it is extremely rare indeed to have the universal self be the dominant self-model through which a person perceives the world. It is the skin-encapsulated model which is by far the prevalent model. Yet, as we shall see shortly, it is this very model which lies at the root of much of humanity's problems today. To understand just how deeply rooted it can be, let us look at how this model develops.

THE DEVELOPMENT OF DUALITY

The newborn baby is aware of the environment but does not appear to differentiate himself from it. He is not aware of himself as a separate entity. As awareness of physical separateness from the mother begins to grow, so does the awareness of separateness from the rest of the environment. According to most psychologists, a true sense of individuality does not develop until simple language has begun to develop (some, such as Jean Piaget, would claim that full identity of the self is not attained until the age of seven or eight.) This feeling of individuality is reinforced by most languages. The subject-object relationship inherent in their noun-verb structure implies that the actor and the action are quite separate and distinct. This becomes manifest in the growing child's subtle but important shift from 'John wants the ball' to 'I want the ball'. The child is beginning to be conscious of an internal self.

In addition to learning a dualistic language, the growing child learns from his parents how to think and behave. If the parents are seen to be working on the assumption that 'I' am 'in here', completely separate from the environment 'out there', then the child learns to adopt the same model and begins to develop his own thinking along the same lines. So the skin-encapsulated ego develops.

The sense of discreteness and individual uniqueness given by this model does have considerable value. Biologically speaking, we are very much self-maintaining, self-regulating, self-directing organisms, and the notion of a separate individual 'self' is a symbol of this autonomy. The feeling of uniqueness that comes with a sense of a discrete self allows us to distinguish our own selves from others. In addition, striving to maintain a unique individual self ensures a much higher chance of survival for the physiological organism.

At the psychological level, this sense of individuality provides an inner unity to all thought, feeling, perception and actions: it is 'me' 'in here' who is experiencing and doing. It is this which gives us our sense of 'me-ness'.

When the skin-encapsulated self is taken as the *only* sense of self, however, we end up seeing the world solely in terms of 'me' and 'not-me'. This leads us to feel there is an absolute distinction between ourselves and others. We characterize ourselves through the ways in which we appear to each other, and draw our separate identities from those features which distinguish us from others – from our height, weight, age, sex, nationality, skin colour, clothes, house, car, social status, job, friends, character, personality, thoughts and ideologies. Thus our sense of who we are is derived from our perceptions, experiences, and interactions with the external world.

Yet the self is not really any of these things. A person can be of a different height, weight, age, etc., but this does not make their sense of 'I-ness' different. Thus we derive a sense of self from what we are not.

Deriving our identity in this way is, to borrow an analogy

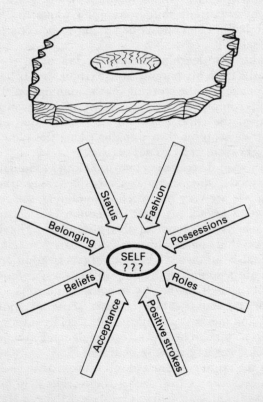

FIGURE 10. A hole in a piece of wood may often be described in terms of the qualities of the wood that surrounds it. Similarly the self, when it is not known in its own right, is usually defined in terms of its surroundings.

from the American philosopher Daniel Cowan, like describing a hole in a piece of wood in terms of the colour, shape and texture of the wood that surrounds the hole (e.g., 'it is a brown, round, smooth hole'). The hole's identity is, so to speak, derived from the wood around it. Most people describe a hole in this way because the qualities of the hole itself are much more abstract; it is easier to describe the qualities of the wood than the transparent air that fills the hole. Similarly, our sense of personal identity is usually derived from what surrounds the self (i.e., from our experience of the world). What lies within is much more difficult to describe.

When an externally derived sense of identity is our only sense of identity, it becomes the most precious of possessions. Without it 'I' would, quite literally, cease to be. (This is a major reason why physical death is so greatly feared; it implies the separation from everything that one has depended upon for a sense of self.) Yet the derived self is as transitory and ephemeral as the experiences from which it is derived. It needs continual maintenance, nurturing and protection, and people will go to the greatest lengths to ensure it gets this sustenance. Much of human activity is geared towards establishing and defending our identities, and the low synergy we observe in society can often be traced back to this need.

TO BE LOVED, TO BELONG, AND TO BELIEVE

Deriving our sense of identity from our interaction with others, we need people to recognize and reaffirm our existence, and frequently spend considerable effort achieving this. Life becomes a search for personal reinforcement (what are often referred to as positive psychological 'strokes'). There is not necessarily anything wrong in this; it certainly does make us feel good. But if strokes are the mainstay of our sense of well-being, the search for them can consume a vast amount of time and energy. Some psychologists estimate that as much as 80% of our interactions with other people come from the need for reinforcement.

At the same time, our vulnerability to emotional injury is very high; the derived self is extremely fragile. Events are seldom seen as neutral, and what is not reinforcing is usually seen as threatening. As a result, time not spent pursuing positive strokes may be spent avoiding negative strokes.

When we do receive negative strokes we feel hurt, and the result is often unhappiness and depression. A study of melancholy by Gerald Klerman published in *Psychology Today* (April, 1979) showed that 'the main forces that initiate depressive responses are threats to the psychosocial integrity of the individual – to the sense of self', and concluded that melancholy and depression were the most common psychological complaints of our times.

Another way in which the derived self handles negative strokes, particularly criticism, is to call up its psychological defence mechanisms, such as justification, blocking, and retaliation – methods the injured identity uses to make itself strong again. But since this reinforcement/protection process is essentially endless, and never fully satisfies the self's hunger for reaffirmation, most people unconsciously use many other tactics to bolster their sense of identity. One of these is gathering possessions.

We acquire possessions to show who we are, to give our 'selves' some status. Personal identity often comes to be measured in terms of material possessions, whether they be houses, cars, TV sets, hi-fi systems, electronic gadgets, paintings, libraries, furniture or whatever. When the status connected to a particular possession drops off, it may be discarded or traded for something with a little more prestige. The need to trade in last year's car for next year's model, for example, is usually born of the need to sustain a sense of who we are, rather than to provide a more satisfactory means of transportation.

Many advertisements prey upon the need to reaffirm a sense of self. Buy a certain model of car and you will be the good looking, super-cool, immaculately dressed owner, admired by everyone on the street. (Or, even if you do not completely fit the bill, you can at least feel you are more that sort of person, you can identify more with that image.)

The derived self, constantly striving to reaffirm its existence, often finds added security by identifying with something larger, such as a group or belief system. Belonging to a particular group, whether it be a social group, political group, religious group or private club provides the derived self with the feeling of 'safety in numbers'. Similarly, people live in the 'right' parts of town, belong to the 'right' clubs, know the 'right' people, drive the 'right' cars, go to the 'right' places for their vacation, wear the 'right' clothes, listen to the 'right' music and even smoke the 'right' cigarettes – not because any of them are necessarily better, but because they support one's identity with a particular group.

Fashion likewise depends on the need to reaffirm one's self. In the early seventies, there was a craze for platform shoes, with as much as four inches of block between the foot and the ground. Such shoes did not catch on because everybody was clamouring for such footwear, but because it was presented as fashion, and to be in fashion was to belong to the 'right' group. (This fashion in particular had the added bonus of making the wearer several inches higher, boosting the ego even further.) Millions of people succumbed to platform shoes, even though they caused countless twisted ankles, injured backs and ruined postures. The derived self evidently considered these disabilities a fair trade-off for the reaffirmation of belonging-ness.

Such behaviour may be relatively harmless to society as a whole, but the need to belong to a group can lead to much more serious problems. As soon as another group appears which threatens one's sense of belongingness, people can change dramatically. When, for example, a group of a different colour moves into a town, otherwise peaceful citizens may suddenly become antagonistic, verbally aggressive and even physically violent. Adam Curle, professsor of Peace Studies at the University of Bradford, points out that this 'belonging-identity is the motive force for xenophobia; for the mindless patriotism of "my country, right or wrong"; for the pseudo-mystical yearning after blood and soil; for the arrogant superiority of the local man over the stranger. It is an attribute which we all finally share, and in doing so contribute to the most dangerous dilemma of the human race.'

Another strong source of identity is our beliefs, which we will go to equally great lengths to defend. When they are questioned, the derived self may feel threatened, often triggering strong emotional reactions as we struggle to preserve the status of our own viewpoint, or attack opposing views. Even when we think we are arguing rationally, we may bring to our aid a number of ingenious devices for proving we are right: selective perception, diversion, appealing to authority, misrepresentation, defamation, blinding with facts or jargon, to name just a few.

But the repercussions of holding onto beliefs as part of one's identity can extend dangerously far. Governments will often doggedly stick to their policies, even though these policies are no longer working, rather than admit their political beliefs might be wrong. And some of the bitterest and bloodiest wars in history have been fought in the defence of belief systems.

THE LOW-SYNERGY SOCIETY

The continual need to sustain and reaffirm a sense of identity derived from experience – whether through the search for reaffirming 'strokes', the roles we play, the groups we belong to, the beliefs we adhere to, or some other process – leads us to use the world to feed the self. This results in an exploitative mode of consciousness. We become exploiters of our surroundings, of other people, and even of our own bodies. It may be a more subtle form of exploitation than that of the rich industrialist exploiting his poorly paid workers, but it is nevertheless exploitation – the rest of the world is used in the service of the identity. This mode of consciousness lies at the core of low synergy in society.

The essence of synergy is that the goals of the individual are supportive of the group's goals. But the need to maintain a derived sense of identity is often in conflict with the best interests of other people and those of the group as a whole. Like a child, the derived self needs immediate gratification, and this inevitably leads to the sacrifice of long-term goals in the pursuit of short-term benefits – the antithesis of synergetic behaviour.

One example of low synergy is people topping up their fuel tanks at the first murmur of a cutback in oil supply. The individual need is to avoid running out of fuel, so everyone takes in a few extra gallons, and the excess load on the supply system results in all the filling stations running dry, creating a very real fuel crisis. The individual's action is clearly not in the group interest.

This conflict between the short-term needs of the individual and the long-term goals of the group as a whole leads to what ecologists often refer to as the 'problem of the commons'. The commons were originally the common pastures used for grazing, but the term is now used to refer to other jointly owned resources such as the oceans and the atmosphere. The problem of the commons occurs when people are taking resources out of the pool faster than they are replenishing them – harvesting whales, for example, faster than they can reproduce. It might be in the short-term interest of any individual or group to grasp what he can as fast as he can, but it is clearly not in the long-term interest of everyone else. It will lead to the ultimate collapse of that resource, with nothing more left for anyone, including the original taker.

To try to solve the problem of the commons, psychologists have experimented with various simulation games. Julian Edney, a British environmental psychologist working in this area at Yale University, found that when people experience the simulated downfall of the commons many times, they can learn to curtail individual needs in order to ensure their long-term survival, but only gradually. In the real world, however, we cannot afford to see the food supplies, fossil fuels and other common resources collapse even once, let alone a hundred times. We cannot afford to learn by experience.

The indications are that, in the not-too-distant future, the more developed nations are going to have to cut back on their consumption of oil and other raw materials. Yet, so long as personal identity is strongly bolstered by consumer goods and material luxuries, people are not going to take kindly to the idea of a less consumptive lifestyle. Most of us *understand* that we have to reduce our oil consumption, but without strict rationing or exorbitant price increases, few people are voluntarily going to cut their fuel consumption whilst they still derive a strong sense of identity from driving their own particular car. It would seem that our need for ego-support leads us to resist the changes we most need.

This dilemma is related to what is sometimes called the 'free-rider problem'. This arises when a person perceives himself as a separate independent individual, taking the attitude that 'what I do will not have any effect on the collective'. A person may for example, decide to avoid paying part of his taxes. The loss to the government may be less than a millionth of one per cent, and neither the government nor anyone else in the nation is going to know the difference. The tax avoider not only gains financially, but also continues to enjoy the benefit of all the public services financed by the tax money of others.

The lack of any obvious solution to this problem has been used as an argument for coercive legislation, to force people to forsake their individual goals. But this only patches over the real conflict. As long as a person's dominant need is to take care of the self, he will only go along with the collective goals because he fears for his own welfare, or because he is concerned about the esteem of his fellows.

In some senses, the free-rider is right: he personally does benefit and no one individual suffers as a result of his actions. Yet if everybody followed this argument it would be disastrous.

The distinction between the self and the world leads ultimately to a 'you or me' approach to life, both on a personal and national level. People vie with each other to stay top dog; scientists keep their work secret so they can be the first to publish; high-rise housing goes empty while people go homeless because the situation is in the interest of the owner; catholics and protestants blow each other up because they cannot live together; nations fight over resources because they cannot share them; and rich countries hoard grain while others fight famine, because it's in their own economic and political interest.

Some people might argue that 'the system' is to blame for many of these problems. But when today's crises are viewed in the light of the derived self, it seems just as likely that the shortcomings of society reflect the state of consciousness of the people who compose it, and that the self creates the system rather than the other way around.

EGONOMICS

Two hundred years ago the philosopher and political economist Adam Smith realized that the drive to maintain personal security was the basic force behind capitalism. He argued that the individual, by looking after his own interests, would be 'led by an invisible hand . . . without intending it, to advance the interest of the society.' His theories depicted a society that would naturally be high in synergy.

Unfortunately, capitalist society has not turned out entirely as Smith theorised. He assumed that individuals would act in their own long-term interests. What he did not realise was that when activity is dominated by the need to reaffirm an identity, this derived self seeks short-term or even immediate reinforcement, rather than long-term benefit. Consequently individuals may often act against their own real interests and against those of society as a whole.

But capitalism is not the only economic system that suffers from the weaknesses of the derived sense of self; communism too is subject to it. Whereas capitalism seems to have surrendered to the needs of the ego, communism (and I am speaking here of the present brand of communism practised in the USSR and most East European countries) appears to have made the opposite error – failing to take the needs of the self into account.

Even though the political theory might hold that each person be supported according to his needs, the most urgent personal needs – the needs of the identity – are derided. The communist system has been made to work only through suppression, censorship, and coercion. Collective action and national support are maintained through propaganda which, like capitalist advertising, preys upon the need to reaffirm an identity. Psychological sustenance is still gained through belongingness – belonging to the 'right' type of state.

Yet even in the USSR money buys prestige, and so money remains an important incentive. At present about 3% of the farms in Russia are privately owned, yet this 3% accounts for a far greater proportion of the agricultural production, principally because of the greater financial incentives involved. As a result, the country is beginning to increase the number of private farms, however difficult this may be to reconcile with its political philosophy.

The growing tide of consumerism in communist countries, which so worries the authorities, is likewise only to be expected. A recent article in the Soviet daily *Izvestia* complained of the disease of 'unrestrained acquisitiveness'. It reported apartments stuffed full of non-essential trifles, especially antiques or anything of rarity, a near mania for black-market hi-fi, and couples saving to buy new cars and then snubbing their car-less neighbours. To us in the West these stories may have a familiar ring. And, so long as the self needs reinforcement, this behaviour is not suddenly going to disappear, whatever the economic model under which people live.

As with capitalism, the ideals behind communism may be fine. The present shortcomings lie in people not yet being sufficiently developed within themselves to live those ideals naturally, free from constraint. So long as personal identity is still one of a localised separate self, the communist system is unlikely to lead to high synergy.

Present day communist societies are no higher in synergy than capitalist ones. In neither case are the needs of the individual in tune with the needs of society as a whole. In a capitalist state it is the needs of the individual which dominate and the society as a whole which stands to lose (and ultimately the individual as well), whereas in the communist state it is the needs of society which dominate and the individual who stands to lose (and ultimately the state as well). Regardless of the system, the personal reality is still one of me 'in here' and the world 'out there', of 'me versus you'.

HUMANITY VERSUS NATURE

This dangerous separation of ourselves from others, so symptomatic of the low synergy society, has led us to an even deeper schism – the 'I versus It' approach to the world, prominent in Western culture as 'Humanity versus Nature'. This approach has been strongly reinforced by scientific and technological models which see humanity as the supreme life-form, able to exercise control over the world and tame it to its own ends. Yet it is not science and technology themselves which are to blame for our present critical situation, but the way in which they have been used. In most cases they serve individual, corporate and national egos, rather than humanity and the planet. Nations cannot be more synergistic than the individuals who comprise them, and so they too fall victim to limited and short-sighted goals.

Britain pours sulphur dioxide into the skies to fall as acid rain on Scandinavia. Lead from petrol consumed in California ends up in the Eskimos' food. We steadily tear down giant forests, removing the planet's prime source of oxygen and destroying the natural habitat of over a million species; such behaviour satisfies our short-term needs for fuel, paper and building materials, but does not take into account what will happen when the trees run out, nor the long-term damage we could be doing to the biosphere. Countries continue to plan the building of nuclear reactors because these offer a solution to our immediate energy requirements, even though no one can guarantee that a major accident will not contaminate the planet. We leave a legacy of potential disaster just because it suits our immediate aims.

When things go wrong and Nature fails to follow our plan, we devise a 'fix' to patch over the flaws in our approach so that we can continue onward. European agribusiness, knowing better than Nature, 'fixed' the inefficiency of the old rotation system by becoming intensive, ripping out hedges and cramming in crops. They then 'fixed' the ecosystem they had damaged with extra fertilisers. New hybrids were developed to raise crop yields, yet these now rely heavily on oil for the special fertilisers and pesticides on which their survival depends. It now takes about fifty times as much energy (mostly oil) to produce the food we eat as we get from eating it. What will happen when the oil runs out? Probably another technological fix – although we should not count on it.

A NEW WORLD-VIEW

For nearly every problem facing humanity, we have the

knowledge necessary to change course and avoid catastrophe; or, if we do not yet have it, we know how to proceed in order to gain it. We have, for example, the knowledge and most of the technology whereby we could, over a ten-year period, make the shift from fossil fuels to renewable resources such as hydro-electricity, tidal energy, wind-power, geothermal energy and solar energy, to satisfy the major part of the world's energy requirements. Yet the proportion of the developed nations' energy budget spent on research and development of renewable sources is less than one per cent of that spent on furthering our dependence on the rapidly dwindling supplies of oil.

The real problem lies not in the physical constraints imposed by the external world, but in the constraints of our own minds. The currently predominant world-view is of man the dominator and manipulator of nature, inherently aggressive and nationalistic, with the principal goals of productivity, material progress, economic efficiency and growth. Science is seen as the supreme approach to knowledge, ultimately able to explain all, using technology as the means to achieve anything that is desired.

Although almost universally accepted by the developed nations, this world-view has only been with us for the last few centuries – it emerged with the Industrial Revolution, as a shift away from the predominantly ecclesiastical model of the Middle Ages, which saw religious teachings rather than science as the source of knowledge, and God as the supreme arbitrator. Valuable though it may have been, our current model is clearly no longer working. Too much is going wrong for us to ignore the implications. The old model is losing its usefulness, and is now threatening our continued existence on this planet. Moreover, the longer we cling to the old world-view, the greater the chances that we will end up where we are currently headed.

If we are to avert a collective catastrophe, it is increasingly apparent that some major and fundamental changes will be necessary: changes in the way we relate to ourselves, our bodies and surroundings; changes in our needs; changes in the demands we make of others and of the planet; and changes in our awareness and appreciation of the world. As numerous people have pointed out, a new world-view is needed – one that is holistic, non-exploitative, ecologically sound, long term, global, peaceful, humane and co-operative. We need to shift to a truly global perspective in which the individual, the society and the planet are all given full recognition. In other words, we must shift from a world-view which is low in synergy to one that is high in synergy.

It has been suggested here that the root cause of much of the low synergy in contemporary society is our using the skin-encapsulated model of the self for the mainstay of our identity. Until recently, there had been little reason to question this dualist model; it seemed to work fairly satisfactorily, and most languages and cultural traditions strongly supported it. But the gravity of the rapidly approaching global crises are now forcing us to realise that it contains some fundamental flaws.

To change the global situation, far more than a series of social, scientific and technological paradigm shifts is called for. To shift from a low to a high synergy society is going to require a shift in metaparadigm – a profound shift in our basic self-model. Such a change in consciousness has now become an evolutionary imperative.

This is not to imply that we must rid ourselves of the skin-encapsulated model. We are very much unique biological organisms, perceiving and acting upon the environment, with strong motivations to protect and nourish this individuality.

Yet this is only one side of the self. Spiritual teachers, mystics. and visionaries have repeatedly affirmed, whatever their culture or time, that we are more than just biological organisms bounded by the skin. We are also unbounded, part of a greater wholeness, united with the rest of the Universe. This is the other side of our identity, and it is this aspect of the self which is needed to balance the sense of individuality and separateness. It is to the nature and the means of unfolding this deeper identity that we shall now turn.

CHAPTER 8

The quest for unity

Yes, there is That which is the end of understanding.
Yes, there is That which you will only understand
At the mind's flowering time,
When she shall leaf and bud and burst
Into her fullest inflorescence of fine flowers.
But should you try to trammel up the mind
And bind her, confine her, and *strive* to turn her inwards
You will not understand.

For there is a power of the mind's prime
Which, rising like the sun in all his majesty,
Shines forth with rays of thought at one with feeling.
Hold still the vision of thy Soul in purity
Freed from all else.
Let but thy little mind be empty of all things else;
All things save one – thine aim
To reach the end of understanding,
For that subsists beyond the mind.

The Chaldean Oracle (Anon.)

Deriving our sense of self from our experience is, we saw, like describing a hole in a piece of wood in terms of the qualities of the wood, rather than those of the hole itself. Our true nature, like the air which fills the hole, is much more difficult to describe. It is far less tangible than the derived self, yet it is the one common element to all experience.

The search for this underlying self, the true essence of our identities, has engrossed the attention of people the world over for thousands of years. The eighteenth-century Scottish philosopher David Hume, for example, repeatedly looked within himself in an attempt to discover something he could call his true self. Yet he found that 'when I enter most intimately into what I call *myself*, I always stumble on some particular perception or other, of heat or cold, light or shade, love or hatred, pain or pleasure. I never catch *myself* at any time without a perception, and never can observe anything but a perception.' There is, he concluded, no experience of the self, and, since he believed all ideas came from experiences, there could be no idea of the self.

Hume's difficulty is not an unusual one. It is born of the fact that the self that underlies all experience cannot itself be experienced in a tangible way. For us to have any kind of experience two components are necessary. There must be an object of experience – that which is being experienced – whether it be something in the external world, a sensation in the body, a mental image or a feeling. And there must be a subject of experience, an experienc*er* – that which is having the experience. It was this subject of experience that Hume was looking for in his search for the self.

Yet to be able to experience the self in the same way that anything else is experienced would require that the self become an object of experience. This is essentially why such an approach fails to find anything. It is rather like being in a room where the only light comes from a torch on one's head and then proceeding to search for the source of the light. All that one sees are various objects, reflecting the light, and giving it quality and form.

This is not to say that the subject of all experience is completely unknowable, only that this Self (with a capital 'S' to distinguish it from the derived self of the previous chapter) cannot be experienced in the same way that anything else can be experienced, for it is itself the experiencer. It is not an experience of 'I am this' or 'I am that', but the 'I' which experiences 'this' or 'that'.

Thus whatever we may think or perceive our selves to be is not the pure Self, the subject of all experience, but the self as it appears within our consciousness. This is why in everyday experience we come to identify with the contents of our awareness, with our experience, rather than with our true selves.

To 'identify' means literally 'to make the same' (from the Latin *idem-ficare*). What we identify with is what we make the same as 'I', yet the very fact we have to *make* those things the same as 'I' implies they are not really 'I' to begin with. You can see this for yourself by taking an aspect of yourself that you feel is intrinsically you and asking if the feeling of 'I-ness' would be different if that aspect were different. You might start by asking if the feeling of 'I-ness' would be different if you were three inches taller; or if you were of a different race; or if you were the opposite sex; or if your body was in any other way different. You would certainly appear different, but the very fact that *you* could have different physical characteristics shows that the 'you' who has those characteristics has not changed. You have a body, but you are not your body.

The same applies to your emotions. One day you may feel one thing; the next day you may feel something different. Again the intrinsic sense of 'I-ness' has not changed, only your emotions have changed. You have emotions, but you are not your emotions. The same is also true of thoughts and desires. You have thoughts, but you are not your thoughts. You have desires, but you are not your desires.

What then is the nature of the pure Self? What is the essence of this sense of 'I-ness'?

THE PURE SELF

We have seen that for the Self to experience the Self, the experiencer would also have to become the object of experience. In this situation, there would no longer be any distinction between subject and object, and no room for any interaction between them. Experience, as we normally know it, would cease. This would be a state of pure consciousness, devoid of all content.

How are we to understand a state in which there is nothing to be conscious of and yet consciousness itself remains? A good analogy might be the distinction between hearing and listening. Wherever you are, if you listen you will usually hear various sounds around you. But suppose you were in a completely silent room. You could still listen, but there would be nothing to be heard. Likewise, in a state of complete mental silence, you, the experiencer, are conscious, but there is nothing to be conscious of. You are – but you are not any thing. It is a state of pure *being*.

Since there is no content to this state of consciousness, there is no means by which a person might distinguish his own most intimate appreciation of himself from anyone else's. One is, in effect, in touch with a universal level of the self. If there is any identity at all in this state, it is of an at-one-ness with humanity and the whole of creation.

For most people such a state of consciousness occurs rarely, if ever. Generally, our attention is directed outward, into the world of sensory experience, away from the pure Self. Even when the attention is directed inward, it is usually still preoccupied with thoughts of one kind or another. To not have a thought in one's head – not even the idea 'I do not have a thought in my head' – is a very rare thing.

Although not part of most people's common experience, there is abundant testimony that this Self-awareness is possible. Descriptions of such states crop up again and again in the writings of mystics and religious teachers around the world, and from all periods of history. Yet because the pure Self has none of the usual properties attributable to an experience, it becomes very difficult to describe in words. Indeed, the very act of describing it makes it an object of experience rather than the subject. This is why many mystics speak of it as the 'ineffable' – that of which one cannot speak. It lies beyond descriptions, beyond any idea we might have, and attempts to describe it inevitably make it an idea of some kind.

The Mundaka Upanishad, an ancient Indian spiritual text concerned very much with the nature of the pure Self, contains an admirable summing up of this difficulty:

It is not outer awareness,
It is not inner awareness,

Nor is it suspension of awareness.
It is not knowing,
It is not unknowing,
Nor is it knowingness itself.
It can neither be seen nor understood,
It cannot be given boundaries.
It is ineffable and beyond thought.
It is indefinable.
It is known only through becoming it.

In a similar vein, the ancient Chinese text, the *Tao Te Ching*, speaking of the true nature of all things (the Tao), opens with:

The Tao that can be told is not the eternal Tao.

Yet clearly it is not of much value to remain completely silent about the nature of this Self. So, bearing in mind these difficulties, let us look at how some of the mystics – those engaged in a personal quest for union with the divine or sacred – have referred to this state, and look in particular at the way in which it has led them to an immediate awareness of their oneness with the whole of creation.

Chuang-Tzu, a Chinese mystic living in the fourth century BC, wrote quite simply that in this state:

I and all things in the Universe are one.

Plotinus, the third century Egyptian philosopher said:

Man as he now is has ceased to be the All. But when he ceases to be an individual, he raises himself again and penetrates the whole world.

Meister Eckhart, the thirteenth century Christian mystic, noted his experience that:

All that man has here externally in multiplicity is intrinsically One. Here all blades of grass, wood and stone, all things are One. This is the deepest depth . . .

And Henry Suso, a German Dominican, wrote:

> All creatures . . . are the same life, the same essence, the same power, the same one and nothing less.

What these mystics seem to be saying is that, 'At the deepest level of my being I am of the same essence as you, and the same as the rest of the Universe. We are all of this same essence, and I experience my Self as such. This is what we are "the same as", this is our deepest level of identity.'

If you find this is difficult to understand, do not be too perturbed. For most of us, accustomed to thinking of ourselves as completely separate individuals, it is indeed a very difficult concept to grasp. An analogy used by several different spiritual teachers might help make the idea clearer. Our individual consciousnesses are like drops of water taken from an ocean: each drop is a unique individual drop, with its own particular qualities and identity; yet each drop is also of the same essence as the ocean – water.

THE PERENNIAL PHILOSOPHY

This experience of unity can be seen to form the core of all mystical and religious traditions. On the surface, the various religions might appear to offer very different teachings about the nature of reality and the means towards achieving salvation or liberation. But once one begins to pare away cultural trappings and the additions and corrections imposed by later commentators and translators, a basic teaching begins to emerge which is common to them – we are, at our cores, united.

The writer and novelist Aldous Huxley, who studied the the major religious and mystical traditions in considerable depth, referred to this basic teaching as the *Perennial Philosophy*. It is, in Huxley's words, that which 'recognises a divine Reality substantial to the world of things and lives and minds; the psychology that finds in the soul something similar to, or even identical with, divine Reality; the ethic that places man's final end in the knowledge of the immanent and transcendent Ground of all being.'

The Upanishads, for example, tell us that:

What is within us is also without.
What is without us is also within.

A statement remarkably similar to a passage in the recently discovered Gospel according to Thomas:

The Kingdom is within you and it is without you.

And *The Awakening of Faith*, an early Buddhist treatise written by Ashvaghosha, declares similarly that:

All things from the beginning are in their nature Being itself.

An important point to note about the perennial philosophy is that it is not a philosophy in the Western sense, for it is not an ideology or belief system. Rather, it is based on the experiences of those of who have tasted such states. It is not so much a set of ideas to be thought about or debated, as an invitation to turn within and discover these truths for oneself. The consequent changes in awareness, life-style and morality may be profound, but they come as a result of knowing this state of pure Being, rather than through the acceptance of any conceptual system or doctrine.

Moreover, the perennial philosophy repeatedly declares that the realization of our essential oneness is not reserved for a select few. Because the Self is common to everyone, we all have the potential to be aware of our real inner nature. We thus find similar statements coming from secular people as much as from the traditionally religious. Here, for example, is the poet Tennyson:

Individuality itself seemed to dissolve and fade away into boundless being, and this was not a confused state but the clearest, the surest of the sure, utterly beyond words – where death was almost a laughable impossibility – the loss of personality (if so it were) seeming no extinction but the only true life.

Edward Carpentier, the nineteenth century social scientist and poet wrote:

> If you inhibiṭ thought (and persevere) you come at length to a region of consciousness below or behind thought . . . and a realisation of an altogether vaster self than that to which we are accustomed. And since the ordinary consciousness, with which we are concerned in daily life, is before all things founded on the little local self . . . it follows that to pass out of that is to die to the ordinary self and the ordinary world.
>
> It is to die in the ordinary sense, but in another, it is to wake up and find that the 'I', one's real, most intimate self, pervades the Universe and all other beings – that the mountains and the sea and the stars are a part of one's body and that one's soul is in touch with the souls of all creatures.

To someone who has not experienced such states, this sense of inter-connectedness might seem a bit far-fetched. But the notion of a unifying element within all forms of manifestation is not just a philosophic concept. Within the last fifty years, this idea has been gaining increasing support from the seemingly-unrelated field of modern physics.

THE MYSTICAL PHYSICIST

There is one obvious way in which we are all of the same essence. You, and I, and everything else in this Universe, are composed of a selection of a hundred or so different types of atoms, which are themselves composed of a few elementary particles (electrons, neutrons, protons, etc.) These various particles, it now appears, may in turn be composed of just three even more elementary particles called quarks. In this respect we are all of the same physical essence. This, of course, is a very simple form of unity. The unity that the mystic is talking of, however, is a far more profound oneness – a oneness of consciousness and matter, a oneness on the level of experience.

It was Albert Einstein's famous 'Special Theory of Relativity' published in 1905, which first suggested that we may be mistaken to think we can completely isolate the experiencer from the physical world which is being experienced. This work contained the revolutionary idea that what appears to us as time and space are not absolutely fixed. For example, if two people travelling at different speeds are viewing the same events, then what one person measures to be the distances and times between the events will be slightly different from the measurements of the second person. This is the case even after they have taken into account their differing speeds. Einstein showed that, although they may appear to us as very separate phenemona, time and space are but different aspects of the same thing (the space-time continuum). But the oneness of time and space is broken whenever we observe the Universe, and different observers may see different proportions of space and time. In short, Einstein showed that the motion of the observer partly determines how reality is perceived.

Twenty years later another German physicist, Werner Heisenberg, working in the field of atomic physics, put forward his Uncertainty Principle. This showed that it was impossible to measure both a particle's position and its speed beyond a certain limit of accuracy. The more accurately you measured one aspect, the less accurately could you measure the other. The very act of accurately measuring a particle's position would make its velocity indeterminate (or conversely, an accurate determination of its velocity would make it impossible to ascertain its exact position). Heisenberg had therefore shown that the act of observation itself affects that which is being observed.

For physicists of the time, who regarded the observer and the observed as separate detached entities, these two conclusions had shattering implications. Somehow the mental and physical worlds were interdependent.

Since then, theoretical physics has paid considerable attention to the way in which the many varied phenomena of the Universe appear to be but manifestations of a single underlying whole. Einstein had shown that not only were time and space one, but so were the electric and magnetic forces, and energy and matter (his famous equation $E = mc^2$). He spent the latter years of his life seeking for a yet grander unity, a Unified Field Theory, which would show the four fundamental forces of physics (gravity, the electro-magnetic force, and the so-called 'weak' and 'strong' nuclear forces) to be but different manifestations of one single principle. Despite the years he spent devoted to this problem the Unified Field continued to elude him.

Subsequent work in physics, however, is now showing that his hunch was probably right. The physics involved is very complex and I will not attempt to explain it here, but developments in what is known as 'Gauge Theory' suggest that the four basic forces of the physical Universe can indeed be thought of as manifestations of a single process. Gravity and electricity may appear very different to us, with very different physical characteristics, yet physicists now believe that they are but different aspects of one force.

Another finding of modern physics suggests that we may be mistaken to think of atomic particles as separate and distinct entities. The notion of separateness is useful at the everyday level, and has in the past been a valuable model for understanding atomic structure, but it may not be the ultimate truth. At the most fundamental level there seem to be only patterns of energy giving rise to the *appearance* of separate particles. This suggestion carries with it the revolutionary implication that we are all intrinsically interwoven into the fabric of the Universe, and are in some respects interconnected, even though appearing physically separate.

One approach to understanding this interconnectedness has come from the British physicist David Bohm, who has introduced the notion of implicate order. Whereas explicate order is the Universe we see around us (the cause and effect world described by the various laws of physics), implicate order is a level of order not perceivable by the senses or by any physical apparatus. At the level of implicate order, every part of the Universe contains the whole Universe enfolded within it. This is a strange and difficult notion to grasp, and a helpful analogy can be found in the new photographic technique of holography.

In a normal photograph, each point on the photograph is a particular part of the final image, and for the image to be seen correctly all the points need to be in the correct position. In a hologram, on the other hand, each point on the photographic plate records data about the whole image. Every part of the image is encoded in every part of the plate. Looking at a hologram with the naked eye, however, you see nothing but a very fine pattern of ripples or waves. But when light of a particular kind is shone through the plate, the image can be made to appear, and it stands out from the plate as a three dimensional image. Since any region of the plate contains information about the whole image, it can give rise to the whole image (although the smaller the region of plate used, the fuzzier will the image be). In this sense the image is 'enfolded' throughout the plate.

Bohm's theory of implicate order suggests that the physical Universe may be like a hologram, with the whole of space and time somehow encoded in every part of it. This implicate order is never perceived directly. What we see is the explicate order – specific forms which are generated from the underlying implicate order. Ultimately, concludes Bohm, the entire Universe has to be understood as a single undivided whole in which separate and independent parts have no fundamental status.

Another British physicist, Richàrd Prosser, has proposed a physical explanation of how such enfolding could occur. (Again the detailed physics need not concern us – the mathematics alone needed considerable computer analysis before the theory was validated.) Essentially he has suggested that what we consider to be a single, isolated, elementary particle can be considered to be an infinite wave pattern spreading out in all directions across the Universe. The nature of the waves is such that they cancel out everywhere except in one very tiny region, and it is in that region that we find the 'particle'. This means that everything is in a sense everywhere, but only appears, or manifests, at one particular point.

Even though these theories are coming from physicists, they are beginning to sound more and more like the teachings of the mystics. Indeed, if the Universe is ultimately a unity we should expect such a convergence of ideas. The physicist is probing the deepest levels of objective existence using the tools of physical experimentation, reason and mathematics. Whereas the mystic is probing the deepest levels of subjective existence through personal introspection. If they are independently approaching an ultimate unity in which both the physical and mental worlds converge, we should not be surprised if their revelations and realizations to sound more and more alike.

In short, physics is discovering the Perennial Philosophy for itself. It is reaffirming that at the deepest levels we are all one, and that mystics and visionaries may well be people who, in one way or another, have obtained a direct cognition of the nature of reality.

CHAPTER 9

Awakening the self

A human being is a part of the whole, called by us 'Universe', a part limited in time and space. He experiences himself, his thoughts and feelings as something separated from the rest – a kind of optical delusion of his consciousness. This delusion is a kind of prison for us, restricting us to our personal desires and to affection for a few persons nearest to us.

Our task must be to free ourselves from this prison by widening our circle of compassion to embrace all living creatures and the whole nature in its beauty.

Albert Einstein

The perennial philosophy repeatedly affirms that we are all ultimately one, that this oneness is knowable as the pure Self at the very core of our being, and that this realization is open to all; indeed, it is our birthright. That the vast majority of people do not live in such a state of consciousness is because we have all been conditioned – some would say hypnotised – by our upbringing and culture to see only the superficial side of our identity. The pure, universal Self is ever present, but most of us are 'asleep' to it. As the eighteenth-century English poet, painter and visionary William Blake wrote in the *Marriage of Heaven and Hell*:

If the doors of perception were cleansed, everything would appear to man as it is, infinite.

For man has closed himself up, 'till he sees all things thro'
narrow chinks in his cavern.

As an analogy, let us return for a moment to the example of
the black and white patches on page 100. With appropriate
visual clues you probably came to see them as a picture of a
man's face. Once you have seen the face it is usually 'obvious'
that face was, in a sense, always there. Without a shift in set,
however, it may not be seen at all – 'What is he talking
about? A face in there?'

It seems to be the same with the pure Self. The reality of the
self as a universal Self may be there, 'obvious' to those who
know it, but if we have not made the necessary shift in set we
simply cannot see it. This is what the Zen Buddhist masters
have repeatedly claimed: 'You are already enlightened; all
you have to do is wake up to the fact'.

The question is: How do we wake up? How do we
dehypnotise ourselves? Answering this question is one of the
most crucial tasks now facing humanity.

It is no good knowing intellectually that we are inseparable
from the rest of the Universe, if reality is still perceived
dualistically, as me 'in here' and the rest of the world 'out
there'. So long as this dualistic mode of consciousness
dominates, we shall continue to misuse valuable resources,
misapply technology, mistreat the environment, mismanage
our lives, and remain on the path towards collective disaster.
Any new laws, policy changes, social reforms, will still fall
prey to the personal reality of an isolated self – however much
we may argue to the contrary. Rather than building a truly
holistic world-view, we will only succeed in re-modelling the
structures and policies existing within the old model, and the
intrinsically exploitative mode of consciousness will continue
to condition our thoughts and actions.

For humanity to accomplish a profound shift in attitude,
the skin-encapsulated model of the self needs to be augmented
by the *realization* that the individual is an integral part of
Nature, no more isolated from the environment than a cell in
the body is isolated from the human organism. The word
'realization' is crucial. Many people already have the intellectual knowledge that we are inseparable from the rest of

Nature, yet it is becoming increasingly clear that this understanding is insufficient to bring about a truly radical change in the way we treat one another or the planet.

A truly holistic ecological ethic cannot be built into our attitudes, policies and actions unless it is first built into our selves. It needs to be an immediately experienced fact of life, an unavoidable premise of all our thoughts, perceptions, feelings and actions. We need to realize our essential oneness with Nature, not just with our intellect and reason, but with our feelings and with our souls. It must become an undeniable part of our reality.

From what we have already seen of the egocentric model, it might appear very difficult to make such a shift. This model has been built up from early infancy; it is one of our strongest conditionings. It is further reinforced by language, social institutions and the behaviour of those around us. Moreover, a self-model cannot be changed by thinking, by argument, by analysis, or by simply deciding to change it, since it is the frame of reference that underlies all thought, argument, analysis and decision making, and as such is beyond their scope.

As we saw earlier, our self-model conditions all our mental activity, and in this respect may be considered as a metaparadigm – like a scientific paradigm, but pervading all areas of thinking. Scientific paradigms are likewise very resistant to change; yet change they do. We might therefore find some clues as to how to initiate a shift in metaparadigm – that is, in our self-model – by looking in more detail at how parallel shifts occur in science.

THE COPERNICAN REVOLUTION

One of the classic examples of a paradigm shift is the Copernican revolution in astronomy. The old paradigm of the Earth at the centre of the Universe, with the Moon, Sun, planets and stars revolving around it, had been formulated by the Greek astronomer Ptolemy around 140 A.D. This paradigm was based on Plato's belief that circular motion was the perfect motion. Heavenly bodies, it was held, displayed perfect motion, and must therefore move around the Earth in circles. Observation showed, however, that the planets did not move in smooth circles around the Earth; their speed varied and on occasion they even reversed their motions. Ptolemy managed to account for these anomalies by assuming that the planets moved round smaller circles whose centres moved round the Earth in circular orbits. This produced a curve known as an epicycle, which approximated the path of the planets (see Fig.11), and allowed the principle of circular motion to be retained.

As measurements became more accurate, more anomalies were discovered, and these were accounted for by the addition of more complex epicycles and the introduction of various oscillations until the system became very complex indeed. This model, cumbersome as it was, survived virtually unchallenged for 1300 years.

Over the centuries a few brave souls had suggested that the Sun, not the Earth, lay at the centre of the system. This theory had even been put forward by some of the early Greek astronomers but they had received little recognition. It was only in the sixteenth century that Nicolaus Copernicus attracted more attention to the idea by giving it a clear mathematical formulation. He showed that if the Sun lay at the centre, many of the anomalous motions of the planets could be explained at one stroke. Copernicus, however, still adhered to Plato's idea of perfect circular motion and still accounted for irregularities in terms of epicycles.

The suggestion that the Earth did not lie at the centre of the Universe was heresy to the Church, and Copernicus refused to publish his work till late in his life. His fears were well founded. Some of his supporters were punished by the Church, some even burnt at the stake. And when his own work was eventually published, it was placed on the papal index of forbidden books.

The next step towards a new paradigm came eighty years later when Johannes Kepler, a German astronomer, had the good fortune to come into possession of volumes of extraordinarily accurate astronomical observations made by the Danish astronomer Tycho Brahe. Working on this data, Kepler came to the realisation that the sun-centred system could explain the various motions, without any need for complex 'epicycling', provided that the planets travelled in ellipses rather than circles.

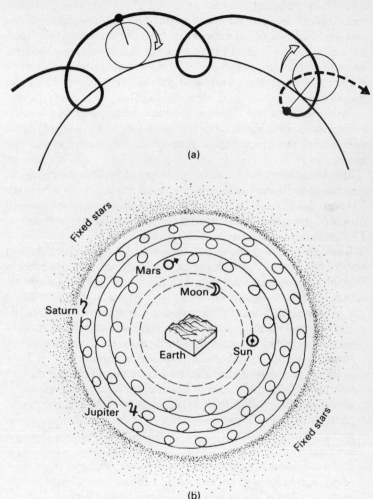

(a)

(b)

FIGURE 11. (a) An epicycle: the curve drawn out by a point attached to a circle which is rolling around a second circle.

(b) The Ptolemian view the Universe with the planets moving around the Earth on epicycles.

These two major shifts – the shift away from the idea of Earth being at the centre of the Universe, and the shift away from circular motion – together gave birth to a new paradigm, a radically different world-view.

Yet even though the model worked very satisfactorily, it was still not readily accepted by the 'establishment'. When, for example, the Italian mathematician Galileo, using the new invention of the telescope, gathered evidence to support Kepler's model, the professors in the universities felt very threatened. They united against Galileo, and secretly denounced him to the Church for blasphemy. He was hauled before the Inquisition, where he was forced to 'abjure, curse and detest' the absurd view that the Earth moves round the Sun.

It was not until Sir Isaac Newton published his major work, the *Principia*, in 1687, that the new model was finally accepted. Newton laid down the basic laws of gravitation, which provided the theoretical underpinning for Kepler's theories. The paradigm shift was now complete.

Looking at the general patterns that characterise this and many other paradigm shifts, science historians have shown that such changes usually go through the following stages:

1. Anomalous findings that cannot be explained in terms of the currently accepted paradigm. Initially these anomalies may be rejected as spurious or fallacious findings, or the model may be 'stretched' to incorporate them.
2. An increase in the number of such anomalies until they can no longer be so easily discounted or accommodated, and it is realised that the paradigm may be at fault rather than the observations.
3. The formulation of a new paradigm that explains the new findings.
4. A transition period in which the new paradigm is challenged by the establishment, sometimes leading to bitter struggles by those who are attached to the old paradigm.
5. Acceptance of the new paradigm as it explains further observations and predicts new findings.

THE IDENTITY SHIFT

Like a scientific paradigm, the skin-encapsulated model of the self gains status through its ability to provide a coherent framework for experience. This status is reinforced by the fact that most of what happens to us can be incorporated within the model of me 'in here' and the world 'out there'.

Every perception of the outside world fits into the egocentric model, precisely because it is an experience of the outside world. As far as normal experience is concerned, therefore, there would seem to be no anomalous phenomena that might threaten the model. Even the impending global catastrophe, which has its roots in the dualist self-model, is still perceived by me 'in here'. Today's crises are crises only for the social, economic, technological and political paradigms; they may make us intellectually aware that something is wrong with our world-view, but they do not directly challenge our experience of a skin-encapsulated self. Consequently we tend to ask what we can do about the world, rather than what we can do about our selves, and the self-model remains unquestioned.

The one phenomenon that directly challenges the skin-encapsulated model is the personal experience of unboundedness, of oneness with the rest of creation – the immediate awareness that 'I' and everything else are, at their most fundamental levels, of one essence. This direct personal experience of unity is the one anomalous observation that cannot be incorporated within the skin-encapsulated model. This is the crisis for the old identity, revealing the model's incompleteness and starting the shift towards a new self-model.

In the case of a scientific paradigm, one single anomalous observation does not by itself produce a major shift; it is usually either ignored or explained away. The same is likely to occur with shifts in identity. There are many instances of people who have at times tasted such states of unity, when suddenly they perceive themselves and the world as a single whole. These states of consciousness may be brought on by a beautiful sunset, long-distance running, meditation, drugs, intense emotion, the view of planet Earth from space, or seemingly by nothing in particular.

But it is one thing to have had such experiences; it is quite another thing to have this awareness of unity become the fundamental basis of all perception, thinking and action. Most people who experience such unitary states find that afterwards they return to the dualist, skin-encapsulated model of the self. The *memory* of having experienced an intimate oneness with creation may well remain, but the oneness itself is no longer an inescapable reality. From the standpoint of the old self such experiences may, like the anomalies in a scientific paradigm, be cast off as mental aberrations, hallucinations, or some strange quirk in brain function.

For a true shift in identity to begin, the anomalies must build up until a point is reached where the old egocentric identity is no longer tenable and begins to lose its status. This implies that the experience of oneness usually needs to be repeated again and again before it can begin to be included as part of one's personal reality. The identity has to be reconditioned to unity. This, as we shall see in the next chapter, is the purpose of many spiritual disciplines and practices of meditation. They are processes through which one can come to know this other reality, and partake of this awareness repeatedly.

THE NEW COPERNICAN REVOLUTION

In the case of paradigm shifts, one of two things can happen to the old model. It can be rejected as erroneous, as was the case with the Copernican revolution. Alternatively, the old model can be incorporated within the new model. This is what happened with Einstein's revolution in physics, where Newton's laws of motion were retained as a special case of the Theory of Relativity.

It is this second type of shift that is needed with our model of the self. We should not be trying to reject the skin-encapsulated model, because the sense of individual uniqueness and separatedness does have important values in terms of our biological identity and autonomy. The many problems and inappropriate behaviours that we looked at earlier stem not from this egocentric model of the self, but from our dependence on this as the *only* form of self. The shift in identity should therefore be one that can accommodate the skin-encapsulated model as a valuable, but partial, view of the self.

Willis Harman, a futurist at the Stanford Research Institute, has referred to this shift of consciousness as the 'New Copernican Revolution'. In the original Copernican Revolution, the geocentric model of the physical Universe was turned inside out. The Earth lost its position at the centre of the Universe; the Sun became the centre with the Earth revolving around it. In the New Copernican Revolution, our egocentric model of the material world is similarly inverted. The individual ego, which has for so long been assumed to be the centre of our inner universes, would be put in its proper position, that is, revolving around the pure Self, the true centre of all consciousness – T.S. Eliot's 'Still point of the turning world'.

The original Copernican Revolution came about by gaining greater knowledge of the whole system, not by changing the movement of the Earth, nor by trying to hold the sun still. So the New Copernican Revolution will come about by expanding our awareness of our own inner natures, not by trying to destroy or restrain the ego, nor by trying to hold onto an idea of the pure Self. To try either to destroy the individual identity, or to think as if one were at one with the rest of creation, without first gaining a personal experience of the fact, would only create dissonance between theory and experience. Moreover, a strong sense of an individual self is probably essential for social interaction, communication and self-improvement. To destroy the ego would be to take away 'the motor of the world'. We would be reduced to aimless vegetables.

The state in which the pure Self has become a permanently established personal reality is what many spiritual traditions refer to as self-realisation or enlightenment. Used in this sense, enlightenment means more than being particularly wise, aware or well-balanced; it denotes a clearly defined state of consciousness. The enlightened person is still functioning as an individual organism, preserving a sense of biological autonomy. And to this awareness of individuality is now added the equally real awareness of unity with the rest of creation. Oneness and separateness become two different perspectives of the identity.

The important point about the enlightened state of consciousness is that the individual is no longer dependent upon the environment for his sense of personal existence. The pure Self can in no way be affected by the ups and downs of the outside world. Thus the incessant need to repair an injured ego, and re-assert it whenever some threat arises, no longer exists. Indeed, to reaffirm the ego at the expense of others and the world around would now be in direct conflict with the experience of oneness with the external world.

The fact that the enlightened person no longer needs to derive a sense of self from his interactions with the external world does not mean that he is devoid of personality, character or idiosyncrasies. In these respects enlightened people are as individual as anybody else – as any survey of enlightened mystics and religious teachers will instantly show. What is important is that they are no longer psychologically attached to these attributes. It is in this sense that they are 'ego-less'. It is not that they have lost their individual egos; it is that they have lost the continual need to reaffirm them. Action ceases to be dominated by the ego and becomes more appropriate to the situation at hand.

Since he is no longer psychologically dependent on his experience, the enlightened person is not kicked around by the world. Personal criticisms, loss of a job, or other events which before would have been a source of anguish are still very real, but they are no longer perceived as personal threats. This is borne out by the experience of many people, who, though not enlightened, are nevertheless already making progress in this direction. They often remark that it is not so much the sense of oneness that first becomes noticeable, but the experience of being at ease with themselves, combined with an increasing sense of inner security and invulnerability.

With this liberation from the needs of the derived self comes a simultaneous opening of the heart. Love, which before was conditional – conditional upon the person finding another person in some way loveable – now becomes less conditional. The enlightened person begins to experience a spontaneous love for every creature and every thing, whatever their qualities or attributes.

The growth of such a love is a recurrent theme in most religions. It is, for example, a principal aspect of the Christian tradition. The Latin word *caritas*, sometimes translated in the gospels as charity, originally meant 'dearness' or 'closeness' – love in the sense of an at-oneness. This is love in a much deeper and far-reaching sense than just friendliness or looking after one's neighbour. Thus we take the phrase 'Love thy neighbour as thyself' as an injunction to feel and act in a certain way. Yet it was probably never meant as an injunction at all. It is not an attitude we must try to have, but the description of a state of consciousness we need to reach – the state in which you know your neighbour (and everybody else) to be of the same essence 'as thy self'.

For most of us, the experience of true affinity seldom encompasses more than a few other individuals. Rather than feeling an affinity for the rest of the world, most people feel alienated from it. A genuine love for the rest of creation comes from the personal experience of oneness with the rest of creation; the awareness that, at the deepest level, the Self and the world are one. A deep affinity with everyone and everything then arises quite spontaneously, as a natural consequence of being in this state of consciousness.

We find the same message stated very succinctly in the ancient Chinese *Tao Te Ching*:

> Love the world as your own self then you can truly care for all things.

Two and a half thousand years later, Teilhard de Chardin elaborating on this theme, wrote:

> 'It is not a *tête-a-tête* or a *corps-a-corps* that we need; it is a heart-to-heart . . . If the synthesis of the Spirit is to be brought about in its entirety (and this is the only possible definition of progress), it can only be done, in the last resort, through the meeting, *centre to centre*, of human units, such as can only be realised in the universal, mutual love'.

He goes on to add that 'there is but one possible way in which human elements, innumerably diverse by nature, can love one another: it is by knowing themselves all to be centred on a single 'super-centre' common to all.'

The enlightened person experiences a deep and universal compassion, and his life usually becomes one of service, not just service to humanity but to the whole world. In the words of a Buddhist scripture,

> 'The fair tree of thought that knows no duality bears the flower and fruit of compassion, and its name is service of others'.

The enlightened person knows a reality that lies beyond the everyday duality of 'me' and 'not me' and the suffering it causes, and his compassion for humanity makes him want to help others achieve this realisation. For this reason many Buddhist teachings have proclaimed that the enlightened being does not rest till he has seen the enlightenment of all beings.

It is towards this goal of the enlightenment of all that humanity must now move. Those who have achieved enlightenment have generally been few and far between. But, if the world is to be transformed and a high synergy society is to become a reality, such a shift in consciousness will need to be widespread.

HEALING THE PLANETARY CANCER

A worldwide shift towards a higher state of consciousness has important implications for the hypothesis that human society is like a planetary cancer. We saw in Chapter One that there were a number of parallels between the way a malignant growth develops in the human being, eventually destroying the body on which it is ultimately dependent, and the way in which humanity appears to be eating its way indiscriminately across the surface of the planet, disrupting and possibly destroying its planetary host. In malignant tissue the individual cells cease to function as part of a larger organism. They feed and reproduce themselves at the expense of the rest of the body. They are in a sense egocentric cells. Cancer is, in this respect, a low-synergy phenomenon.

The synergy of a living organism depends upon the DNA molecules in the nucleus of each of its cells. The precise arrangement of numerous smaller molecules along any particular strand of DNA gives rise to a particular genetic code. Different people carry slightly different genetic codes on the DNA, but within any particular individual the same genetic information is contained in virtually every single cell. This coding determines not only the specific characteristics of the individual, but also the way in which every single cell functions. Most importantly, it provides an essential element of common programming which links the individual cell to the organism as a whole.

The reason why a particular cell becomes cancerous is that this genetic code is in some way disturbed or interfered with. This can happen in several ways. It may be caused by the effect of toxic chemicals, by exposure to X-rays or nuclear radiation, or simply because, in the process of constantly regenerating its billions of cells, the body occasionally makes an imperfect one.

In the healthy organism the odd imperfection is quickly dealt with, but in the weak or stressed organism the disturbance may result in a cell that is missing some of the information that ties it back to the system as a whole. Without the necessary links to the rest of the system, the relationship of the rogue cell to the whole is one of low synergy, and the cell begins functioning in a low-synergy manner. It may also start producing other cells in which the genetic code is similarly disturbed, leading to a malignant growth.

At the level of society, that which ties the individual back to the system as a whole, and which therefore performs the function parallel to that of the gene, is the pure Self. As the genes in the DNA provide the common programming essential for the high synergy of the cells in the body, so awareness of the Self provides the common programming essential for the high synergy of the 'cells' in the social super-organism. As with the gene in the cell, the pure Self is an unchanging entity. It is at the centre of all consciousness, and is the same throughout society, just as the genes are the same throughout a particular individual.

It may well be that human cancer and the planetary cancer are even more closely related than this analogy suggests; they may be two different symptoms of the same problem. Human cancer has steadily become more prevalent in the last few decades, particularly in the more developed Western nations, at the same time as those nations have become more malignant in their approach to the environment. Some of the increase may be due, paradoxically, to better health care; when tuberculosis was a major killer, not so many people had the chance to go on to develop cancer. But modern lifestyles are also a major element. Contemporary diets have been implicated, as have general attitudes to life. We are also discovering that numerous products (e.g. hair dyes, suntan oils, asbestos fibre, photocopying fluids and chlorinated drinking water) may be cancer inducing, not to mention numerous airborne pollutants and radioactivity. These are all examples that stem from the low-synergy approach of contemporary society, and which seem to lead to low synergy and cancer within the individual.

In order to reverse this malignant trend in society we need to be tied back once again to the system as a whole through an experience of our oneness with the world. Interestingly, the Latin word for 'to tie again' is 're-ligare', and the word 'religion' originally meant just this; that which ties us again to our common source. This does not mean to imply that we need to return to conventional religion, for, as the next chapter will show, convential religion has generally lost the art of 'religare'. What we do need is a spiritual renewal, a widespread shift in consciousness along the lines experienced by the great mystics and proponents of the perennial philosophy.

Such a shift has now become an evolutionary imperative, not only for the well-being of both individuals and society as a whole, but also for Gaia herself; it is the path to a spontaneous global remission. In this respect the person whose goal is self-realization, whether he be a yogi in a Himalayan cave or an office worker in London, is helping to change the world at the most fundamental level. Such people are perhaps the ultimate revolutionaries.

EVOLUTION FROM THE INSIDE

We have seen that the shift from an ego-dominated model of the self to a more universal model seems to be a very necessary ingredient in the development of higher synergy, and the transformation of humanity into a healthy social super-organism. In this respect the development of self-realization is supporting the general thrust of evolution towards the progressive integration of the human species.

We can, however, go further than this. We can see the development of higher states of consciousness to be an essential part of the evolutionary process itself.

The previous evolutionary leap to self-reflective consciousness not only allowed us to be aware of ourselves as conscious, thinking beings; it also gave us the capacity to be conscious of the essence of consciousness itself, the pure Self. It thereby bestowed upon us the possibility of becoming spiritually enlightened.

Furthermore, with the emergence of self-reflective consciousness the platform of evolution moved up from life to consciousness. Consciousness became the spearhead of evolution. For the first time on Earth evolution became internalised. Thus the urge that many people feel to grow and develop inwardly is nothing less than the force of evolution manifesting within our own consciousnesses. It is the Universe evolving through us.

This inner evolution is not an aside to the overall process of evolution. Inner, conscious evolution is the particular phase of evolution that we, in this little corner of the universe, are currently passing through.

From this perspective the movement towards a social super-organism and the mystical urge to know an inner unity are complementary aspects of the same single process – the thrust of evolution towards higher degrees of wholeness. To flow with evolution is, therefore, to explore our own inner selves, and find unity and wholeness within us.

The question now facing humanity is how can we facilitate this inner evolution, and, even more important, can we do it in time?

CHAPTER 10

The spiritual renaissance

When the Tao is lost, there is goodness.
When goodness is lost, there is kindness.
When kindness is lost, there is justice.
When justice is lost, there is ritual.
Now ritual is the husk of faith and loyalty,
   and the beginning of confusion.

Tao Te Ching

The experience of unity with the whole of creation, in addition
to being closely connected with spiritual and religious traditions, is an experience that has received considerable attention
from a number of psychologists.

Some of the first psychologists to look seriously at religious
experience were William James and Carl Jung, followed in the
1950s by Abraham Maslow, Roberto Assagioli, and others.
Out of this work, a new school of psychology – transpersonal
psychology – developed in the late 1960s; its main focus has
been the study of religious and related experiences. Previously, psychotherapy had been preoccupied with treating people
with mental or emotional problems. But psychologists such as
Maslow turned away from the study of the sick to look at the
mentally healthy, and at the exceptionally healthy in particular. Maslow found that such people had a high incidence of
'peak-experiences' – states in which they felt 'at one with the
world, really belonging to it, instead of being outside looking

in . . . the feeling that they had really seen the ultimate truth'. They had a 'sense of the unity of everything, and of the universe itself being alive'. In this respect such experiences sound very much like glimpses of the unitive Self.

Maslow observed that these people were composed, stable, and integrated members of society. They also displayed a characteristic he termed 'self-actualisation' – 'the actualization of potential, capacities, and talents, as fulfilment of mission, as a fuller knowledge of and acceptance of the person's own intrinsic nature, as an unceasing trend towards unity, integration or synergy . . . '. Most importantly, self-actualisers tended to be centred on problems external to themselves rather than on the task of ego-maintenance. They had a strong sense of identity with humanity as a whole, and a feeling of belonging to something bigger, even the whole of creation. From such descriptions it would seem that these people were probably moving along the path towards what we have called enlightenment.

A recent study by two American sociologists, Adam Greeley and William McCready, showed that peak experiences are far from uncommon. Forty three per cent of the people they interviewed had had an experience of going beyond their normal self, 20% on more than one occasion. Most of those interviewed had never spoken of their experience to anyone before, generally because they were afraid that they might be laughed at; yet the fact that so many people had had such experiences would suggest that many of the people in whom they might have confided would have had similar experiences themselves.

Common to many of these experiences were feelings of joy and happiness, an inability to describe the experience in words, an awareness of the unity in everything, and a realisation that this is the way things really are – again, all characteristics of the unitive Self. Moreover, as with Maslow's work, Greeley and McCready found a strong relationship between these experiences and psychological well-being.

Studies such as these go a long way towards supporting the perennial philosophy's claim that such states can be known by anyone, and are not the prerogative of a select few. However, it is one thing to have tasted such experiences; it is quite another to have had them so frequently that they become the dominant mode of consciousness. This brings us to the question of what, if anything, we can do to facilitate these experiences and make them much more commonplace.

The traditional answer to this question has been to look to conventional religion. In all there are dozens of different religions in the world, and many times that number of particular sects (Buddhism alone contains over 600 different sects). Yet, separate and unique as each religion might appear to be, it is possible to see a common theme underlying them all. Walter Stace, one of the most eminent philosophers of religion, studied at length the writings and teachings of the great religious teachers, and came to the conclusion that the central core of all the major religions was the experience of oneness with creation. In other words, they each appeared to be based on the perennial philosophy.

Each particular tradition originally arose from the teachings of individuals (Christ, Buddha, Moses, Mohammed, Mani, Zoroaster, Guru Nanak, Shankara, Lao Tse, etc.,) and a close examination of what they said, or were reported to have said, suggests that they were each in their own terms referring to this basic unitary experience. Christ may have spoken of the Kingdom of Heaven, Buddha of Nirvana (deliverance), and Shankara of Moksha (liberation), but in doing so they each appear to have been describing aspects of enlightenment. Moreover, if one looks at the practices they taught, whether they were prayer, meditation, devotion, abstention, dancing or prostration, they each appear to have been giving prescriptions whereby the ordinary person could come closer to that state.

Once the teacher had gone, however, his teachings began to become distorted. This is inevitable: it is the equivalent of entropy in the field of knowledge. Each time a message is passed on from one person to another there is some slight change; something may be inadvertently omitted or some little extra thing included. It is rather like taking a photocopy of a photocopy of a photocopy; with each copy the image becomes progressively more blurred. Similarly spiritual teachings have inevitably become distorted as they were passed down. The medium destroyed the message.

As far as the theoretical aspects of a teaching are concerned, distortion can be minimised by writing down or memorising the philosophy and doctrine. But the actual techniques and practices are much more delicate, and often cannot be neatly put into words. Most spiritual practices require guidance from an experienced teacher, and only a slight distortion or misunderstanding can cause a technique to lose its effectiveness. When this happens adherents to a particular tradition become cut off from the goal of the practice: the state of unity consciousness. The net result is that the means to the experience of oneness are lost much more rapidly than the descriptions of the state.

Conventional religion today reflects the tragedies of this continued differential distortion. Doctrines and dogmas abound, and their adherents argue endlessly over which ones are the best. Yet, without the means to experience the states of consciousness being discussed, true enlightenment remains an unattainable dream for all but a lucky few. The experience of unity with the whole of creation may have been their aim, but the major religions today do not facilitate this experience – they have become but the fossils of enlightenment.

What humanity urgently needs today are the means to bring about a widespread shift in consciousness. This will come about not through a revival of any particular religion, but through a revival of the techniques and experiences that once gave these teachings life and effectiveness. We need to rediscover the practices that directly enable the experience of the pure Self and facilitate its permanent integration into our lives.

PATHS OF AWAKENING

Such a revival is already underway. Throughout the Western world there is a rapidly growing number of spiritual masters and gurus, teaching different meditation techniques and paths to enlightenment. There are also a growing number of therapies and training programmes, all aimed ultimately at bringing about an awareness of the inner Self. Whether or not they are all effective in this respect is a question we shall consider shortly. They are, however, indicative of an increasing trend.

A large number of these practices involve some form of meditation, though the term is used to mean many different types of technique. Underlying nearly all meditation practices is the basic premise that to contact the underlying Self, the mind must be cleared of its normal clutter of sensory input and endless trains of thought.

Even when sitting quietly doing nothing in particular, most people find there is some internal dialogue occupying their attention to a greater or lesser extent. As a result, they are not aware of the 'I' who thinks, only aware of what they are thinking about. Thus a common goal of most meditation techniques is to come to a state of consciousness in which there is no thought: a state of complete mental silence. Since in this state all experience (in the normal sense of the word experience) has ceased, only the pure Self, the experiencer, remains.

This is not a blanking out of the mind. It is a common misconception to imagine that one can arrive at this state by deliberately making the mind empty. But usually this only shifts the internal dialogue to the thought that 'my mind is blank' – a deceptive and misleading *thought*. In true meditation one leaves verbal thought behind. The internal dialogue, which seems to occupy so much of our waking consciousness, decreases and eventually disappears. To see the various ways in which different meditations set about achieving this, let us look very briefly at some of the more common approaches.

In Transcendental Meditation (TM), one of the most widespread practices in the West at present, the person sits down quietly and silently repeats a 'mantra', which as far as TM is concerned is just a meaningless sound, although in some other practices the mantra may have a specific meaning. In this meditation one attends to the mantra in a passive manner, not forcing it into any particular form or rhythm. This passive mode of attention is greatly helped by the fact that the mantra has no meaning; it does not, in itself, set you thinking on long trains of associative thought. As with most meditation techniques one shifts from an active 'doing' mode of attention to a passive 'letting be' mode. As a result, the attention is led to experience the normal thinking process at quieter and quieter levels until eventually all mental activity fades away completely.

The normal active mind might be likened to a room full of people, everyone chattering away to each other. The experience of TM corresponds to everybody beginning to talk more and more softly until eventually the room falls completely silent. In this state the listener would still be conscious and listening, although nothing would be heard. In a similar way, when the mind becomes completely still, one is nevertheless conscious, although nothing is being thought. One has effectively transcended normal thinking – hence the name Transcendental Meditation.

A rather different approach to meditation comes from the Indian philosopher-teacher Bhagwan Shree Rajneesh. He holds that the Western mind is too active and too tense to be able to settle down into a silent meditation. His students are, therefore, first encouraged to do a 'dynamic meditation' in which they dance, jump around, and shout in order to get rid of tension. Such intense activity might appear to be the opposite to meditation, but Rajneesh claims that, having given vent to many of their inner tensions, people are better able to settle down to a state of mental silence.

Most types of Buddhist meditation also start with techniques aimed at helping the mind settle down. In some schools, particularly those of Tibetan Buddhism, this often involves visualization exercises, which again take the attention away from verbal thought. Other schools start with breathing exercises, which in addition to having a settling effect in themselves, can also serve as a technique of passive attention, treating the rising and falling breath much like a mantra.

Although the majority of meditation techniques aim at bringing the mind to a state of stillness, some practices take a somewhat different approach. A student practising more advanced Buddhist techniques, for example, may try to realise the essential Self by progressively disidentifying with the more superficial levels of identity, that is, by realising one is not one's body, ideas or feelings. As the process is repeated, uncovering and then disidentifying with subtler and subtler levels of the self, each step takes one nearer to the attributeless pure Self.

In some Zen Buddhist schools the shift in identity is brought on much more dramatically. The master gives the student a *koan*, a paradoxical question or apparently insoluble riddle, such as 'What is the sound of one hand clapping?' Such puzzles are impervious to solution by reason alone – although the student may repeatedly return to the master thinking he has worked out the answer. Eventually, after lengthy pondering and in a state of extreme frustration, when his reasoning and discursive faculties have exhausted themselves, he may suddenly break through into 'satori' – a flash of enlightenment. He has not necessarily found any new interpretation or idea: rather, he has transcended the normal discursive mind and, for a moment, broken through his skin-encapsulated world view.

Such practices are just a few of the many different spiritual approaches to self-realization, and each of those I have mentioned is far more complex than suggested by these very brief descriptions. Even so, the diversity of possible approaches is readily apparent. There are, moreover, numerous other processes and activities that can have similar effects.

Many people have found the state of inner silence coming through physical techniques like hatha yoga and t'ai chi. Some experience it through long-distance running or various other forms of prolonged physical exertion. Some have reached this state through fasting, through pain or suffering, through drugs that modify the brain's functioning, or through sexual experiences. Yet whatever the path, and however deliberate or accidental the breakthrough, the result is nearly always the same: a transcendence of the skin-encapsulated ego and an opening to a deeper unifying self.

Yet, despite this diversity of practices and opportunities, true enlightenment remains a rarity. Why is this?

One reason may be the general level of consciousess of society as a whole (something we shall be looking at in the next chapter). A second reason may be that few of the practices are as effective as they might be. The causes for this are various: The techniques may be difficult to teach correctly, or too easily distorted by the student. Some approaches may require the development of considerable skill before they work. In many cases it may be some time before a person notices any changes in himself, and lacking this feedback he may throw the practice away. In other cases the person may occasionally have clear experiences of higher states of consciousness, motivating him to keep on, but the experiences may not be frequent enough to produce a permanent shift in consciousness. Certain programmes may require a high degree of personal instruction and consultation with a teacher, and there may not be sufficient numbers of such teachers available. The approaches derived from the East are sometimes incompatible with an everyday Western lifestyle. Frequently, the beneficial effects of a particular technique may be negated by other influences in a person's life: use of alcohol or other drugs, fatigue, poor health or stress. And finally, the social environment in which most people live does not usually reinforce any intimations of unity they might have.

If there is one lesson that has been learned since the rush for enlightenment began in the 1960s, it is that enlightenment doesn't follow as rapidly as many of the pundits would have us believe. But this is not necessarily a reason for despondency. As far as the large-scale enlightenment of society is concerned, we are still pioneers – and pioneers make mistakes. (It is, perhaps, worth remembering that the early steam engines that began the Industrial Revolution were not that efficient or successful either.) Yet, as long as the idea of an inner development of consciousness remains alien to externally-oriented ways of thinking, it will be easy to scoff at, or even scorn, the trials and errors of the inner explorer.

THE MARRIAGE OF EAST AND WEST

In order for the shift to a higher state of consciousness to become widespread, society will need to develop techniques or processes that are simple to practise, can be incorporated into most people's day-to-day life, are easily disseminated throughout society, and produce the required shifts in consciousness fairly rapidly. Although most of the techniques available today do not appear to achieve these ideals, it is very likely that science – psychology and physiology in particular – will help us realise these goals.

Just as microscopes, computers, electronic equipment, and a vast range of experimental and analytical techniques have already led to an increased understanding of the outer world, so science and technology are now leading to an increased understanding of the inner world. Electron microscopes, for example, are helping neurophysiologists examine how individual brain cells function and communicate. Advances in computer analysis and electronics are leading to a better understanding of the extremely complex patterns of electrical activity generated by the combined interaction of the billions of cells in the brain, and the ways in which the various regions of the brain interrelate in different states of consciousness. Biochemists are uncovering a wide range of chemical processes that have a direct effect on the brain's functioning and our experience. Other approaches are looking at how different states of consciousness lead to different perceptions of the world around, and are studying which factors, both internal and external, trigger changes in consciousness.

As we begin to combine this growing scientific understanding of the brain and consciousness with the knowledge and techniques of mystics and spiritual teachers, we will be better able to see how the techniques work, how they can be improved or developed, and how best to facilitate the transition from 'experimental steam engine' to 'mass transportation'. With this marriage of East and West will come the birth of a new discipline, the field we might call psychotechnology. More than just the study of the mind or psyche, it will be the application of techniques to improve the functioning of the mind and raise the quality of experience and the level of consciousness.

Progress in this direction is already being made in a number of areas. Over the last decade, science has begun to take a serious interest in meditation and its potential values. In the past decade more than a thousand research papers have been published on the physiological, psychological and biochemical effects of meditation. The general conclusion of the various studies is that meditation produces a physiological state which is the exact opposite to that of stress. The whole system becomes very relaxed, and brain activity shows patterns characteristic of a quiet, restful state of consciousness.

An associated area of research, biofeedback, provides an excellent example of a dawning psychotechnology. In this process a person is given information about aspects of his physiological functioning such as brain rhythms, heart rate, blood pressure or skin temperature. This is usually done by connecting the measuring apparatus to a light or sound which is switched on whenever the physiological process is in a certain state (e.g., whenever blood pressure drops below a certain level). The person is encouraged to make the light or sound come on as much as possible, which he generally finds he can do by adopting certain mental attitudes or images. Through such processes a person can learn to control many physiological processes that Western physiology had generally thought to be beyond voluntary control.

As the techniques of biofeedback become more sophisticated, and as we come to know in more detail what takes place in the brain during deep meditation and in mystical states of consciousness, it is becoming possible to use biofeedback to induce or facilitate such states. One does, of course, need to be careful that the Western techniques do not adversely interfere with the traditional techniques (by dividing the attention, for example), but if the combination proves to be successful, it could lead to a significant acceleration in the process of self-realisation.

Sensory isolation tanks provide another example of the coming together of science and meditation. The purpose of these 'tanks' is to provide an environment in which there is minimal sensory input. Usually the person floats in water at body temperature in a sound-proof and light-proof enclosure – hence the term 'tank'. Such conditions facilitate the withdrawal of the attention from the senses, which is probably why many find that in these environments the inner states characteristic of meditation can be induced more rapidly, and often more deeply.

Hypnosis is another useful tool that has potential for facilitating meditative states of consciousness. Although known about in the West for over a century, hypnosis is still very poorly understood. Nevertheless its profound effects on a person's consciousness are well-established. But rather than just being a tool for curious stage effects, or an alternative to anaesthesia in surgery, the principles of hypnosis, when combined with deep relaxation and visualisation exercises, are now being seen as powerful means of inducing higher states of consciousness, temporarily shifting a person's identity away from the egocentric model to an awareness of the pure Self. Research in this area is still in its infancy, but it may well prove to be a very powerful tool indeed.

It may also be possible to accelerate one's spiritual growth by biochemical means. Many primitive cultures have used extracts from various herbs, cacti, mushrooms and other plants to bring on altered states of consciousness; and since the 1950s a growing number of Westerners have been experimenting with them. However, most of the chemicals tried so far have tended to produce undesirable mental or physical side effects. Moreover, although the states of consciousness induced may superficially sound like those found in mystical experience, it is not yet clear whether they are in fact the same. Nevertheless, the possibility remains that we may find other chemicals (or synthesize completely new ones) capable of triggering the same brain states as are associated with the shift of identity to the pure Self.

These are just a few examples of the ways in which science may be able to further the development of higher states of consciousness. The field is a rapidly expanding one, and within the next decade or so we will probably see psychotechnology become a major area of scientific exploration, bringing with it developments which we have not yet imagined. Moreover, it will probably be one of the most crucial areas of human activity. If as much money, manpower, energy, time and thought were devoted to the facilitation of higher states of consciousness as have been spent on the arms race, there might be no need for an arms race.

In addition to the development of more effective techniques of enlightenment, there is another very important advance in Western technology that will be of immense value in furthering inner transformation – the communications revolution.

In the past there may have been many teachers who have taught practices of self-realization, but such teachers could only directly influence those in their immediate vicinity. Christ imparted his message to those who lived in his area of the Near East, as did Buddha in northern India, but without the technology of mass-communication, the knowledge and practices had to be passed on from one person to another. This inevitably led to distortion and a loss of effectiveness. This is one reason why no one has yet succeeded in enlightening humanity as a whole, or even a major part of it.

Today, however, we have a variety of means of communication that can be used to make information globally available. Cars, air travel, postal services, telephones, telex, video and audio tapes, satellite links and computer networks have made it possible to communicate with almost anybody in the world in a variety of different ways. In addition, some of these devices enable us to store that information and play it back at later times. With these developments, it has become possible, for the first time in the history of the planet, to spread the means to self-realisation directly and accurately. And curiously, this possibility has come just at the time when humanity as a whole seems desperately to need the shift into a higher state of consciousness. But is this really so curious? Perhaps, from an evolutionary perspective, the ultimate purpose of technology has been to enable society to make this shift.

A NEW AGE DAWNING?

There is a growing number of people who feel that a new human era is dawning, one that will involve a fundamental shift in people's consciousness and their relationship to the rest of the planet. Such thinking is often referred to as the 'New Age' movement. The people who hold these ideas are not unified by any central organisation; rather, they constitute a loose and diverse network of groups united principally by their common attitudes and values. Four recurring themes of the New Age movement are:

- We all have potentials beyond those we are now using, and perhaps beyond those we even dream of;

- Humanity and the environment are a single system;

- We are mistreating and often abusing both ourselves and our surroundings;

- Humanity *can* change for the better.

The New Age movement encompasses a wide and diverse range of interests. There are ecologically orientated groups concerned with the protection of endangered species, organic farming, communal living, alternative technology, voluntary simplicity, energy and resource conservation, nuclear disarmament and other ways in which we can live more in tune with the planet.

There are people and techniques concerned with improving the health and physiological well-being of the individual through jogging, inner sport, the Alexander technique, the Feldenkreis method, bioenergetics, autogenic training, holistic medicine, acupuncture, healing, massage, shiatzu, rolfing, iridology, naturopathy, homeopathy, osteopathy, health foods, whole foods and dozens of different diets.

There are numerous means of therapy including hypnotherapy, hydrotherapy, dreamtherapy, logotherapy, reality therapy, Reichian therapy, gestalt therapy, primal therapy and sex therapy, each concerned with improving psychological health and inner well-being. So also do a variety of other programmes such as rebirthing, biofeedback, enlightenment intensives, sensitivity training, encounter groups, psychosynthesis, neurolinguistic programming, psychodrama, androgeny workshops, actualisations, est and arica.

There are meditations of many different kinds drawn from just about every spiritual tradition, and other practices of a spiritual nature such as t'ai chi, aikido, tantra and yoga.

There are groups interested in developing paranormal abilities such as aura reading, telepathy and past-life experiences. And various forms of divination from astrology and tarot to geomancy and radionics.

In addition, many others are concerned with such areas as holistic education, feminism, natural childbirth and other ways of allowing people to unfold their full potential.

There are magazines entitled *New Age, New Directions, New Humanity, New Equinox, New Times, New Roots*, and *New Realities* (none of these having anything to do with the *New Scientist* group of magazines). There is also a *New Dimensions* radio network. There have been Celebrations of Consciousness, Awakening Festivals, Aquarian Festivals, Festivals of Mind-Body-Spirit, World Symposiums of Humanity, and Omniversal Symposiums.

Many of these groups herald the dawning of a new age – not in the far distant future, but now, in the present. Some find support for this idea in the sheer number of such groups (and those mentioned above are only a small sample) and the rate at which the interest is spreading (and we shall look at this aspect in greater detail in the next chapter). Others cite the astrologer's claim that our planet is moving into a new era – the Age of Aquarius.

Astrology divides the sky up into twelve sections – the twelve signs of the zodiac – and as the Earth moves around the sun each year, the sun appears to move through each of the twelve signs in turn (they talk of the sun being in Taurus, for example). The Earth meanwhile is spinning about its own axis of rotation. This axis is tilted and the tilt points into space in the same direction as we go around the sun (which is why we have seasons – winter in the Northern hemisphere when the North Pole points away from the sun, and vice versa).

But the direction of the Earth's tilt is very gradually changing. This is brought about by the combined attraction of the sun and moon on the bulging equator, though the exact mechanics need not concern us here. The net effect is that the heavens appear to wobble about the Earth, taking approximately 26,000 years to complete a circuit. As a result, our Earthly calendar, based on the seasons, slowly moves backwards through the zodiac (or, to use the technical term, precesses through the aquinoxes), any given date changing signs once every 2,100 years or so.

Astrologers usually take the spring equinox (March 21st) as the beginning of the astrological year, and the position of this equinox in the zodiac determines the characteristic age. For the last 2,100 years or so the spring equinox has been in *Pisces*, but it is now shifting into *Aquarius*. Thus, it is claimed, we are now entering the 'Age of Aquarius' – an age that astrologers characterize as one of increasing harmony, high moral idealism, and spiritual growth. Astrologers are very divided as to exactly when the transition occurred (or if it can even be given an exact date). The general consensus, however, seems to be around the late 1960s with a focussing of votes around 1967.

Whether or not there is anything in such claims is a matter of controversy. Nevertheless, the late sixties, particularly 1967, was certainly a time of transition for many people. It was the heyday of 'flower power', and erstwhile hippies still look back with nostalgia at the 'Summer of '67'. But it was not just a time when people were taking LSD, professing love and peace, and handing out flowers, and when the facades of houses in Notting Hill, Greenwich Village, and Haight Ashbury suddenly sprouted rainbows, flowers and sunbeams. Most of those involved really felt that if everybody in the world were to have experiences of love and unity, then the world could not fail to be a happy and peaceful one. A new age could begin; the formula seemed so easy.

The year 1967 itself ended with the Beatles, at the peak of their career, acclaiming 'All You Need is Love'. The message was simple – and in many ways, correct. Love *is* all you need. If we could love every other person and every other being, then the world would be many times better, if not ideal. But the question remained: How to do it? It is no good simply deciding to love, professing love or acting as if out of love. At its best, this leads to making a mood of love; at its worst it leads to a rather hypocritical self-contradiction.

As we saw in the previous chapter, a genuine, unconditional and universal love stems from the personal experience of being at one with the rest of humanity and the rest of creation. It is this experience that is needed before this vision of an ideal society can be fulfilled.

What happened to society in the 1960s is akin to what happens to an individual during the creative process. Creative breakthroughs usually come after a period of pondering or incubation, but when they do come they come without any warning, as a sudden flash of insight. The essence of a work of art or the solution to a problem is suddenly clear – the path is obvious. Yet the creative flash then has to be implemented, and it may take months or years of work to put the insight into practice.

The late sixties probably represented a similar creative flash for society. Suddenly many people saw how the world could be. The flash, however, did not create the reality, and the task since then has been to implement this insight – finding ways to raise the consciousness of the individual and bring in love and compassion. In this respect the many New Age movements of the seventies and eighties could be seen to represent the multitudinous ways in which people are seeking to translate the vision into reality.

NEW AGE OR OLD

Not everyone, however, sees the pursuit of self-development in such a positive light. Some argue that such a quest has little value for the rest of society, and generally represents a withdrawal from the real issues facing humanity. This is the view taken by Daniel Yankelovich, the social researcher, in his book *New Rules*. He criticises the search for self-fulfillment on the grounds that many people's devotion to inner development is antisocial. The emphasis by Maslow and other humanistic psychologists on the self-fulfillment ethic is, he argues, 'a moral and social absurdity', giving sanction to desires that do not contribute to society's well-being. Yankelovich believes that if Western society is to survive the encroaching crises, we must all make sacrifices and begin to pull together; we must be committed to making society work. Thus, he argues, we must give up 'the inner journey'.

Such arguments rest upon a serious misunderstanding of what the inner journey is all about. Yankelovich interprets self-fulfillment as people having everything they want, 'a career *and* marriage *and* children *and* sexual freedom *and* autonomy *and* being liberal *and* having money *and* choosing nonconformity *and* insisting on social justice *and* enjoying city life *and* country living *and* simplicity *and* graciousness *and* reading *and* good friends, and on and on'. If this were the only type of search for self-fulfillment, he might well be right.

But the pursuits Yankelovich is describing are a far cry from the self-fulfillment argued for by Maslow, other humanistic and transpersonal psychologists, and most spiritual teachers. They are not advocating the satisfaction of self-centred needs – the needs of the derived sense of self to find reaffirmation – but the discovery of the pure self. This is self-fulfillment in a much more profound sense. It is the lack of this much richer type of self-fulfillment that causes people to search for fulfillment in the external world.

We saw earlier that it is through an inner awakening to the real Self that we will we leave behind the ego-centred needs which fuel such hedonism. This kind of self-fulfillment is quite the opposite to what Yankelovich believes it to be, and it is this that society needs the most. It is by tuning in to this deep, unifying level of identity that we will really be able to pull together and make society work.

Other people have argued that, even though the goals of the various self development practices and programmes may be laudable and badly needed, many of the people involved appear to be far from enlightened themselves. They sometimes seem thoroughly egocentric or doctrinaire, as much concerned with their own identity and power as with the enlightenment of others.

Some of those who join a particular programme of self-development may be doing so from a sincere desire to unfold more of their potential or discover their inner Self, but others may come to keep up with the Jones's, to develop promised mental powers, to be part of an in-group, or simply to reinforce their beliefs. People may derive a strong sense of identity from practising a particular system of meditation, ritual or lifestyle, or from being a follower of a particular prophet, guru or leader. Consequently, the more one affirms one's path to be the best path, the more secure the ego feels. Others become attached to the *idea* of enlightenment often resulting in the most dangerous ego-trip of all – proclaiming to all that '*I* have transcended my ego' that '*I* am enlightened'. A rather tragic self-contradiction.

Such attitudes and behaviours might make us feel more than a little despondent about humanity's chances for true spiritual growth. However, they do not necessarily mean that the practices and techniques themselves lead to ego-dominated activity. We are all subject, to varying degrees, to the need to reaffirm our sense of identity, and even those who have become interested and involved in self-development are still going to be prone to this need.

So long as a person has still to realize and integrate the true Self into his life, the means to that goal are inevitably going to be ego-dominated to some degree. Therefore, we should not expect people 'on the path' to behave as if they had reached the goal. But at least these people are putting their energies in the direction of furthering inner growth rather than hunting whales, strip-mining, building arsenals of nuclear weapons or pursuing any other potentially life-destroying activity.

This gives an intriguing perspective on the evolution of consciousness. For many people, the motivation to discover a higher Self springs from those needs that ultimately are to be transcended – the needs of derived sense of self. In effect, these needs serve as the fuel for their own dissolution. Evolution is, so to speak, pulling herself up by her own bootstraps; or as the old adage says, 'using a thorn to remove a thorn'.

CHAPTER 11

On the threshold

Everywhere on Earth, at this moment, in the new spiritual atmosphere created by the idea of evolution, there float, in a state of extreme mutual sensitivity, love of God and faith in the world: the two essential components of the Ultrahuman. These two components are everywhere 'in the air' . . . *sooner or later there will be a chain-reaction.*

Teilhard de Chardin

Another comment often made about the various self-development programmes is that they represent a very minor social phenomenon. Some people might argue that the number of people directly involved in inner growth is very small and, whatever their effects on the individual might be, they are unlikely to have any significant impact on humanity as a whole.

In some respects this criticism is valid; there is little question that at present the development of consciousness is not a widespread human interest. However, if we look at the growth rates in this field it seems possible that this area of activity could be having an extremely significant effect on humanity in the very near future.

Earlier we discussed the exponential curve which characterises most natural growth patterns (page 65). We saw that it was characterised by a constant doubling time (the time it takes for the numbers involved to double) which causes the curve to get steeper and steeper. The human mind seems to find it difficult to handle the accelerating nature of the exponential curve (we boggle at world population growth, for example), and when we make predictions about the future of a given growth curve – particularly when we make spontaneous snap predictions – we may not give this acceleration its full due. If we are not careful, we may forget that the rate of growth will change, and make what is called a linear prediction. Such predictions usually fall far short of the mark.

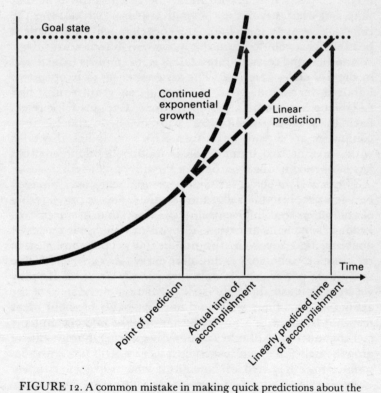

FIGURE 12. A common mistake in making quick predictions about the future of a particular growth is unwittingly to assume a linear growth. If the growth is exponential in character then a given state of affairs will be reached much sooner than predicted.

A good example of such a misjudgement occurred with predictions as to when we could get a man onto the moon. A *Science Digest* article of 1948 declared that, 'Landing and moving around the moon offers so many serious problems for human beings that it may take science another 200 years to lick them.' A few years later a conference of eminent scientists in England came, after lengthy discussion, to a similar though slightly less pessimistic conclusion. They declared that we would not see a man on the moon before the year 2000. The reason they gave was that it would take that long for all the necessary technological advances to be made. They may have predicted accurately according to the rate of growth at that time, but what they did not seem to take into account was the rapid acceleration in technological developments that made it possible for a man to be on the moon only fifteen years later.

Another, and perhaps more familiar, example is to be found in the TV series *Star Trek*. The series was set to happen two hundred years from now, but within ten years reality had caught up with some of its predictions. Captain Kirk refers back to 'primitive computers' with magnetic spools. Such computers are already primitive. Kirk's computer also talks with a synthesized human voice, another development that has happened in one twentieth of the time predicted.

The *Star Trek* script-writers were not being overly naive; very few people in the 1960s foresaw the fantastic proliferation of computers and information technology that has occurred. What has made its growth so difficult to visualise is the short doubling time involved: the number of people employed in information technology is doubling every six years or so, while computing power itself is doubling every year or so. If today we were to make an on-the-spot prediction of the state of the art in ten years time, we would most probably base our ideas on what is happening now, and fail to take into account the fact that not only will it be continually growing, but the rate of growth itself will be increasing just as rapidly. Our 'fantastic' predictions might well fall far short of what is likely to happen. Only when you sit down with paper and pencil, and do the mathematics or plot the curve, can you get a more accurate idea of where the industry is headed.

Rapid as the growth of the information industry is, it may

still not be the fastest growing area of human activity. There are indications that the movement towards the transformation of consciousness is growing even faster. The number of people involved in this area seems to be doubling as rapidly as every 4 years or so. In terms of sheer numbers the movement may not be very significant at present, but it looks as if it might well follow one of the steepest growth curves society has ever seen.

Evidence for this comes from several directions. A study I

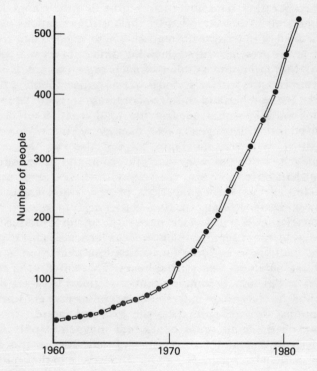

FIGURE 13. The probable growth curve of the number of people involved in self-development and inner growth over the last 20 years. With a sample of only 500 people it is difficult to fit an exact mathematical curve to the data. As it stands the curve is not a true exponential curve; the increase is not smooth and there was a period in the mid-1970s when the growth rate was fairly constant. As a result the doubling time varies, but its average value is just over 4 years.

conducted in England analysed the rate at which people were becoming actively involved in this field. Five hundred people working on inner growth were interviewed. Although a few of them had been doing so for a long time (some as much as 50 years) 40% had started in this direction within the last 4 years. An analysis of the overall rate of growth suggested an average doubling time of just over 4 years. Moreover, this growth rate applies to those who had remained involved; those who may have lost interest were not included.

Corroborating data for such a growth rate comes from membership figures supplied by various organisations active in the field of inner transformation. These suggest that many of them are growing with a doubling time of between 2 and 4 years. Moreover, the number of such organisations is itself growing rapidly, with what looks like a similar doubling time. If both the number and the size of organisations are doubling at this rate, it would suggest the total number of people involved is doubling even faster – once every one or two years. We must, however, take into account the fact that some people may belong to more than one group and so be counted more than once, and also that many of these organizations may initially go through periods of very rapid expansion, followed by a slowing, or even flattening out towards the characteristic 'S' curve (see page 67). *est*, for example, the self-development seminar founded by Werner Erhard experienced doubling every year for its first four years, but is now doubling only once every three years. The net effect of these trends is difficult to compute, but something between three and five years as an overall doubling time seems very likely.

Another way of estimating the growing interest in this subject is the number of books and magazines published. Analysis of the rate of appearance of new titles in this area suggests they have a doubling time somewhere in the region of 3 to 4 years – again supporting the general trend. All in all, therefore, it seems that if we provisionally assume a figure of 4 years for the overall doubling time of interest in this field we are not going to be too far off.

We started this discussion with the criticism that the various consciousness-raising groups were a minor social phenomenon. But from our analysis of exponential growth, we can see that sheer numbers are not so crucial as the overall rates of growth. These movements are growing faster than the information industries, and if the current trends are sustained, the 'consciousness' curve will eventually catch up with and overtake the information curve, however small the former may appear at present. Exactly when the curves will cross depends upon the percentage of the population currently involved in the movement towards higher consciousness. This is difficult to ascertain at present. Government agencies do not yet consider it to be a phenomenon worthy of analysis. A Gallup poll in 1978, however, showed that in the USA the figure may be of the order of two million. In this case the development of consciousness, if it keeps up a doubling time of four years, will occupy the attention of half the US population within the next two decades.

If this seems astonishing it is because once again our on-the-spot predictions fail to do justice to the rapid acceleration of exponential growth. Remember that only twenty years ago the number of people involved in the computer industry was very small indeed – far smaller than the number now working on expanding individual consciousness – yet look what has happened to that curve.

A proportion of those interested in raising consciousness are actually employed in the field (as therapists, meditation guides, etc). If the growth of interest continues to swell, so will the number of people employed in this area, and we may eventually reach a point, possibly sometime early next century, when the employment curve for 'consciousness processing' will overtake that of information processing. The evolution of human consciousness will then have become the dominant area of human activity, and we will have shifted from the Information Age into the Consciousness Age.

This would represent a time when the needs for food, material goods and information were being adequately satisfied, and the major thrust of human activity would move on to exploring our inner frontiers. Self-development would become our prime goal and people would be as familiar with meditation and spiritual experiences as today they are familiar with pocket calculators and cassette tapes. It may sound like science fiction, but it is, I believe, a very possible

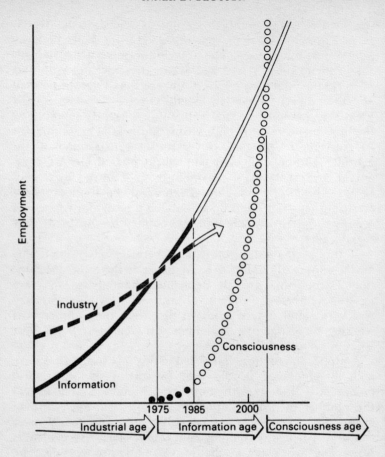

FIGURE 14. Projected growth curve of proportion of population working in the field of inner development (the 'Consciousness Curve'). Although small in numbers, at present, if the current rapid rate of growth is sustained, the number of people working in this area will eventually overtake the information curve.

development. It is a natural extension of the direction in which some of humanity is already headed.

The above arguments have focussed on trends within the

168

West – the USA and Great Britain particularly. They clearly do not apply to the less-developed countries, many of which have not yet even made the transition to the Industrial Age. Nevertheless, it has already been suggested that the lag between the less-developed and the more-developed nations is steadily decreasing, and it should not take as long for the less-developed nations to move into the Industrial Age and from there pass into the Information Age as it did for the West. Similarly, we might expect them to move into the Consciousness Age that much more rapidly. If this occurred, the development of consciousness could well become the dominant human activity over much of the planet within the next century.

In fact, the transition could happen even more quickly than this. First, those who are at present working on inner development are doing so in the context of a predominantly materialistic, externally oriented culture. They are pushing against the inertia of the old consciousness. As the proportion of people reaching higher states of consciousness increases, this inertia will decrease, and at the same time a supportive momentum in the new direction will start building up. The net effect might be that people would begin to find it easier and easier to make progress on the inner path.

The second reason why the transition could come much more rapidly is that we may not have to wait for the majority of a population to be pursuing the transformation of consciousness before we feel the effects. It could be that a small number of people in higher states of consciousness would have a disproportionately positive effect on the rest of society. Such effects could occur if one person's state of consciousness had, in some way, a direct effect on another's. Strange as this notion might seem, it is not totally implausible; indeed, there is growing evidence that it is happening all the time.

MIND LINKS

The idea of minds directly affecting each other takes us into the realm of extra-sensory perception, and telepathy in particular. This has long been a hotly debated topic. The whole field represents a potential crisis for the current scientific paradigm: experiences such as precognition or clairvoyance do not fit the current model of the way the world works; so either the experience must be explained away or rejected, or else the model must be changed. There is not room here to review the current state of the debate – the briefest survey would fill an entire book – and we shall not be immediately concerned with which effects are truly paranormal (i.e., lie outside a normal scientific explanation) and which can be accounted for by orthodox science. Suffice it to say that my own experience makes it extremely hard for me to doubt that something of the kind does occur.

The major emphasis of the research to date has focussed on specific experiences: can one person know what another is thinking or feeling? How accurately can he describe images that another person has in his head? Here we will consider some less specific effects: the extent to which one person's general state of consciousness and overall brain activity may be directly influenced by those of other people.

This is a relatively new field of investigation, having only received serious scientific attention since mid-1970 (principally because the subject had to await the development of the appropriate research techniques and equipment). One of the early series of experiments in the field was conducted by Russell Targ and Harold Puthoff at the Stanford Research Institute. They took pairs of people who already knew each other well and had a degree of emotional affinity (usually close relatives or couples) and placed them in separate rooms at opposite ends of a building, ensuring that they were completely isolated from each other. At randomly selected times, a rapidly flashing light was shone into the eyes of one of the pair. This had the effect of temporarily reducing the overall level of *alpha* activity in the brain. (Alpha activity is a particular electrical brain rhythm usually associated with a relaxed state of consciousness.) Meanwhile, the second person was asked to say when he thought his partner in the other room was experiencing the flashing light. In this task he failed; his scores were no better than those predicted by chance. But although he was not able to 'say' what state the other person was in, his own brain showed reduced alpha-activity at the same time as the first person's did.

Other researchers have since done similar experiments looking at different physiological variables. Instead of producing a specific brain state in the first person, they put him into a stress situation, measuring the stress reaction by changes in skin resistance and other parameters. Again, the second person could not tell when the first person was under stress, but his skin resistance nevertheless changed. The conclusion of these and similar studies is that under certain circumstances we are, in some way, receptive to each others' general state of mind, even though on a conscious level we may be unaware of it.

There is no reason to suppose that such effects are limited to alpha states or states of stress. If the phenomenon is authentic, we should expect it to occur with many other states of consciousness, in particular with the higher states of consciousness associated with enlightenment. And indeed there is considerable testimony to such occurrences in many mystical and spiritual teachings.

An important element of Indian thought is the notion of *darshan* – the belief that an enlightened person can pass on to someone else a taste of enlightenment. Sometimes it may be a touch or a look from the master that confers the experience; sometimes it comes just from being in the presence of an enlightened person. In some traditions the enlightened person or guru need not be there in person, only in the mind.

However *darshan* occurs, the effect is nevertheless very powerful. The experience is often rather like that which comes from an exceptionally deep meditation. Sometimes a five-minute encounter can leave one in a higher state of consciousness for a week, or in some cases for months. And this is not just a feeling of well-being; one actually experiences aspects of the enlightened state.

It could be argued that such occurrences need not involve any direct linkage of minds. They may merely show that, given the right psychological triggers, most of us have the ability to produce such experiences for ourselves. Nevertheless, the experience is very real and can radically change a person's life. Somehow or other enlightenment is contagious.

Similar to this notion is the Christian concept of grace. Christ is said to have brought to man the grace of God, and it is through such grace that man can have a spiritual awakening. The Greek word usually translated as 'grace' is *charis*. A 'charismatic' person originally meant one who bestowed grace, and in the present day there are many charismatic Christian sects based upon the belief that a spiritual experience may be directly bestowed upon a person.

We might hypothesise that similar transferences could occur during meditation. Although not in an enlightened state of consciousness, a meditating person will usually be in a different state of consciousness from the active person, and show different patterns of brain activity. Another individual close by might, then, begin to show some of the same changes in brain activity, even though he may be totally unaware of them. If this second person were also meditating he might well notice some slight deepening or facilitation of his own experience. Moreover, his meditation could have a similar effect upon the first person. Some such mutual feedback process probably underlies the observation of many meditators that they experience more profound meditations when practising in a group – the larger the group, the stronger the effect.

Such phenomena do not just occur within a room. A study conducted in 1979 has shown they can occur between groups separated by great distances. As part of an on-going study of the effects of collective meditation, a group of approximately 3000 meditators gathered in Amherst, Massachusetts. They were all students of Transcendental Meditation practising what is called the 'TM-sidhi program' – an advanced technique that has been found, through a number of studies, to increase the coherence of the brain's activity (coherence, in this sense, being a measure of the extent to which different parts of the brain are functioning in step with each other). The experimenters looked at what effects this group of people, all practising together, could have on another smaller group practising similar techniques a thousand miles away in Fairfield, Iowa. Neither the subjects nor the experimenters knew at what times the other 3000 would be sitting to meditate; yet analysis of the brain activity of the second group during meditation showed an increased coherence *between* individuals whenever the first group was meditating. Coherence between individuals means that the patterns of brain activity of people in the second group were more in tune with each other; their brains became more synchronised.

Just how such effects occur is far from clear, although a variety of theoretical explanations has been put forward. One fascinating theory, which lies within the scope of conventional physics, suggests that during meditation people are setting up resonating electromagnetic waves around the planet, in much the same way as blowing across the top of a bottle can produce resonating sound waves, i.e. a hum. The basis for this hypothesis rests upon the fact that the fundamental resonant frequency for electromagnetic waves (i.e. radio waves) travelling around the Earth (by alternately bouncing off the Earth's surface and the upper atmosphere) is 7.5 Hz (i.e. 7.5 cycles per second). A wave of this frequency, having travelled around the planet, arrives back at its starting point exactly in step with itself, thereby reinforcing itself and setting up a resonance.

It so happens that brain activity during most types of meditation usually includes a strong component at this frequency. Thus, according to this theory, if any faint electromagnetic waves at this frequency were emitted by the brain during meditation, they could set up planet-wide resonances; the planet would 'hum' with the effects of meditation. And the more one resonated with this hum, the deeper the quality of meditation.

The main difficulty with this theory is that any such waves would be exceedingly weak, even with resonance. Although it has also been found that the brain is particularly receptive to waves of this frequency, it is not at all certain that they would be strong enough to be detected.

Another approach to understanding the long-distance transference of higher states of consciousness involves the level of implicate order, the unmanifest level of existence we discussed in Chapter 8. If we are all at some deep level interconnected, and this level of oneness corresponds to the pure Self, then a person deep in meditation, being more in touch with this Self, might so to speak, be sending ripples through the implicate level of existence. And, if other people were also in touch with this level, they might be sensitive to these impulses.

There are many other theories that attempt to explain such phenomena. Some are very complex, involving difficult concepts from modern physics, such as quantum field theory; others lie completely beyond the scope of the current scientific paradigm, and relate such experiences to a sixth sense or to contact with beings on a higher plane.

Which, if any, of such explanations are correct need not concern us here. The important point is that, in some way or other, one person's general state of consciousness appears to set off similar, though generally weaker, effects in other people. This implies that as more and more people in society start experiencing such states of consciousness, other people will gradually pick up the effect, making it easier and easier for them also to reach such states.

Strange as such a process might sound, there are already many instances of such phenomena occurring in nature. In a series of experiments begun at Harvard in 1920 by psychologist William McDougall, rats were being studied to see how quickly they could learn to escape from a maze filled with water. Surprisingly, it was found that successive generations learned the task more rapidly.

Could this be an example of Lamarckian evolution in which parents passed on skills to their offspring? The answer was 'no'. Other researchers in Scotland and Australia found that when they came to repeat the experiments, their first generation of rats, bred from a completely separate strain, started at the same level of expertise as McDougall's *last* generation. Some even 'learned' the task immediately without making a single error; somehow they already 'knew'. Furthermore, as the experiment progressed, successive generations of the control group of rats, who had never been near the water maze, also improved, along with the experimental group. The skill was somehow being learned by other rats, both in the laboratory and across the world.

Rupert Sheldrake, a British plant physiologist, sees this as an example of what he calls 'formative causation'. In his book, *A New Science of Life*, he proposes that systems are regulated not only by the laws known to physical science but also by invisible organising fields – what he calls 'morphogenetic fields' (from the Greek *morphe*, form, and *genesis*, coming-into-being). He sees the regularities of nature to be more like habits than reflections of eternal physical laws. His theory postulates that, if one member of a biological species learns a new behaviour, the morphogenetic field for the species changes, even if very slightly. If the behaviour is repeated for long enough, its 'morphic resonance' builds up and begins to affect the entire species. Thus, in the case of the rats, the more rats who learn the task, the stronger the morphogenetic field becomes, and the more easily other rats learn the task.

As another example of formative causation, he cites the difficulty in crystallizing certain organic compounds that have never been crystallized before. Scientists may work for years until they obtain one crystal. But once this has been achieved, other experimenters across the world usually find it much less difficult to produce their own crystals. And the more crystals that are produced, the easier it becomes to crystallize the compound.

The coventional explanation has been that microscopic 'seed' crystals are carried from one laboratory to another on the beards or clothing of visiting scientists, or by atmospheric currents. But when subsequent crystallizations occur inside sealed containers, as happened with glycerine, this explanation seems to fail. Sheldrake's hypothesis, on the other hand, interprets such phenomena as the building up of a particular morphogenetic field.

Applying Sheldrake's theory to the development of higher states of consciousness, we might predict that the more individuals begin to raise their own level of consciousness, the stronger the morphogenetic field for higher states would become, and the easier it would be for others to experience these states. Society would gather momentum in the direction of enlightenment. Since the rate of growth would now be dependent on the achievements of those who had gone before, we would enter a phase of super-exponential growth. Ultimately this could lead to a chain reaction, in which everyone suddenly started making the transition to a higher level of consciousness.

REACHING CRITICAL MASS

In his book *Lifetide*, British biologist Lyall Watson tells of a monkey tribe on an island near Japan, which had been given supplies of sweet potatoes by researchers studying their feeding behaviour. But the freshly-dug potatoes were covered in dirt and grit, and the monkeys were reluctant to eat them. Then one day, a young monkey dipped her sweet potato in the sea before eating it. She found the cleaner potato much more palatable, and the next day she again dipped her sweet potato in the sea, and continued doing so over the following weeks. One by one the other monkeys began to copy the behaviour and started coming down to the sea to wash their dirty food. Gradually the habit spread from one monkey to another until, according to Watson, it suddenly became universal:

> 'Let us say, for argument's sake, that the number (of potato washers) was 99 and that at 11 o'clock on a Tuesday morning, one further convert was added to the fold in the usual way. But the addition of the hundredth monkey apparently carried the number across some sort of threshold, pushing it through a kind of critical mass, because by that evening almost everyone in the colony was doing it. Not only that, but the habit seems to have jumped natural barriers and to have appeared spontaneously . . . in colonies on other islands and on the mainland in a troop at Takasakiyama.'

If Watson's interpretation is correct then what we are seeing here is an additional phenomenon; not only did the habit spread, it also reached a threshold beyond which it spread as a chain reaction through the society. Could a similar process happen with the development of consciousness? A number of people believe it could.

Teilhard de Chardin, for example, suggested that once sufficient spiritual progress had been made and society had become ripe, then 'it would seem that a single ray of such a light falling like a spark, no matter where, on the Noosphere, would be bound to produce an explosion of such violence that it would almost instantaneously set the face of the Earth ablaze and make it entirely new'.

Moreover, some mystics claim that the number of people

required to set off such a chain reaction need not be that large. For instance, Georges Gurdjieff, the Russian mystic and teacher, said that just 100 fully enlightened people would be sufficient to change the world. Alternatively, if enough people were working on inner development through such techniques as meditation, significant worldwide effects would be felt, even though the people meditating may not necessarily have reached enlightenment. The Maharishi claims that if just 1% of the population were to practice the technique of TM the course of history would be profoundly altered. The 'Age of Enlightenment' could dawn.

In order to see how such a small number could possibly have such a profound effect on the rest of society, it is helpful to consider a parallel phenomenon in the world of physics – the functioning of a laser. Light, from any source, consists of numerous different tiny packets of waves (quanta), each coming from a different atom. In ordinary light these waves are generally all out of step; they are said to be 'out of phase'. If, however, during the brief instant that an atom is about to emit its minute wave-packet, light of a specific frequency (or colour) impinges on it, the atom can be stimulated to emit a light pulse that is in phase with the wave that stimulated it. The new emission thus augments, or amplifies, the passing wave. At low power the net effect is still one of bundles of waves, out of phase with each other, but as the power is increased, a certain level (called the laser threshold) is reached at which a completely new phenomenon occurs: all the little bundles suddenly lock into phase; they are said to become coherent. When they do so, there is a tremendous increase in the intensity of the light produced.

Coherent light increases in intensity because of the different mathematical ways that in-phase and out-of-phase vibrations add up. Waves that are in phase add up as one might expect, a hundred waves acting together being a hundred times as powerful as a single wave. Waves whose phases are only randomly related will, however, partially cancel each other out; they add up only in proportion to the square root of the total number of waves. A hundred out-of-phase waves, for instance, are together only ten times as strong as a single wave. Thus a small number of units acting coherently can easily outshine a much greater number acting incoherently. And the larger the number of units, the more dramatic the effect. Out of one million units, only a thousand (0.1%) would need to act coherently for their effect to dominate.

Scientists working with the Maharishi have attempted to apply similar principles to society in an effort to predict how many people would need to rise to higher levels of consciousness for the rest of society to be affected. Basing their calculations on the assumption that a person who is meditating will not only raise his own level of consciousness, but will have similar effects, however slight, on other people, they have arrived at the figure of 1 per cent as the threshold above which the number of people meditating would have a noticeable effect on a whole city.

This is of course only a theoretical model, and a very simple one; and it is based on rather bold – and to some, dubious – assumptions. Nevertheless, most of the research that has been done on this subject has tended to support the hypothesis.

Preliminary statistical analysis of the crime figures in cities where 1% of the population had learned TM showed that crime rate in these cities fell by an average of 5.7%. In other cities of the same size but with fewer people meditating, the average crime rate rose by about 1.4%. A statistical analysis showed that the probability of this being a purely fortuitous result was about 1 in 200; statistically speaking, this figure was significant and suggested that some relationship might well exist.

Further studies have since been undertaken to investigate whether other factors (such as income, education, unemployment and age) might have been responsible for the change. Could, for example, an increased level of education have been responsible both for the reduction in crime rate and for the growth of interest in meditation? The general finding was that even when factors such as these were taken into account, meditation seemed to remain a significant factor.

If a person meditating does have effects such as these on the rest of society, we may well be headed towards a threshold point, or critical mass of consciousness, beyond which the momentum of rising consciousness would outweigh the inertia of the old ego-based model. If so, crossing the threshold would represent a major transition for humanity. Beyond it society might be completely transformed.

Such sudden transitions are not without evolutionary precedent. We saw earlier that, back at the time of the Big Bang, when the Universe was still super-hot, any matter that did form would have been instantly annihilated. The surrounding heat (disordered energy) was simply too much for the newly created packet of highly ordered energy. Matter only came into permanent being once the temperature had dropped sufficiently (i.e., the general order had increased). Once the conditions were right, however, matter came into being very suddenly. Later, in the primordial soup, life was initially destroyed as fast as it was created. The level of disorder in the surroundings again swamped the highly organised molecular arrangements. Only when a sufficient mass of living systems had been created could life take a permanent hold.

This seems to be a general evolutionary trend, and we could expect that, at this next step in evolution, the new phenomenon of enlightenment would initially appear and disappear many times, being at first swamped by the prevailing low level of consciousness. Only when the social 'atmosphere' had reached a sufficient level of order and organization (i.e. when higher states were sufficiently widespread) could enlightenment become permanently established.

This is probably the main reason why enlightenment has been such a rare thing in the past: society, as a whole, has simply not been ready for it. In this respect Christ, Buddha, Moses, Mohammed and all the other great masters were before their time.

The many spiritual teachers who have appeared over the last few thousand years could be compared with the first bubbles of steam that begin to appear in water as it nears its boiling point. At first, it is not hot enough for these early bubbles of steam to be sustained, and they are rapidly reabsorbed back into the water. They are but the heralds of steam. But when the boiling point is reached, there is sufficient energy for them all to fly free and the water hurriedly turns to steam.

In a similar way, the insights and teachings of the great

masters have been distorted and lost once the teachers themselves died; the wisdom was, so to speak, reabsorbed by the prevailing level of spiritual ignorance. Today, however, the simultaneous convergence of a number of trends could change this. The potential marriage of science and mysticism, the growth of highly efficient methods for disseminating spiritual wisdom, the burgeoning interest in inner development, and the possibility of direct transference of higher states of consciousness, are all combining to make it possible, for the first time in human history, for the wisdom of the perennial philosophy to take a firm and lasting hold.

We may be rapidly approaching a time when the 'bubbles' of enlightenment will no longer be reabsorbed, but will fly free as the whole of humanity begins its great transition. Suddenly everybody will begin to become rishis, roshis, saints, and buddhas.

Furthermore, this transition would be occurring at the same time as the rapid accelerations in many areas of human endeavour are pointing to a major evolutionary transition; at the same time as the population approximates the crucial size of 10^{10}; and at the same time as the connectivity within the human race is reaching a similar complexity to that found in the human brain.

It does indeed seem possible that we alive today could witness the emergence of a true high-synergy society – a healthy social super-organism. If so, we may truly be one of the most privileged generations ever to have lived.

CHAPTER 12

Towards the high synergy society

The world is now too dangerous
for anything less than Utopia

Buckminster Fuller

How would the growth of enlightenment in the individual affect society? How would our values change? What would life be like if we did make the evolutionary leap to a high-synergy society? These are some of the questions we shall be considering here.

One of the first points to note is that it will be humanity as a whole which will make the transition. The most significant changes, therefore, would come in the behaviour of society rather than the individual. Consider again the analogy of water coming to the boil. Before the boiling point is reached the water molecules behave collectively as a liquid. But, as the water boils and turns to steam, the behaviour of the molecules changes radically – they behave collectively as a gas rather than a liquid. The individual molecules, however, have not changed at all, and the laws of quantum physics that apply to each molecule have not changed. It is the relationships between them that have changed, giving rise to a totally different collective behaviour, obeying totally different physical laws. There has been what physicists call 'a change of state'.

We might expect similar transformations with a major

'change of state' in society. The laws of physics, chemistry and biology may not change dramatically, and each person would continue functioning as an individual biological being; we would still breathe, eat, drink, work, play and make love. The most significant changes would happen at the collective level, as our changed relationships, both with ourselves and with other people, began to give rise to a totally different society. It is the 'laws' of economics, politics, and sociology which would be radically changed – laws which are dependent upon the collective behaviour of many people. They would be as different from current 'laws' as the behaviour of steam is different from that of water.

Indeed, so different would our collective behaviour be from the way it is today, any predictions we make may well fall far short of the mark – can a water molecule who has only known life as a liquid ever guess how things will be after the transformation to steam? Nevertheless, even if we cannot do full justice to a high synergy society, some exploration of the general direction in which we could be moving will be valuable, for, as we shall see later, our image of the future plays a role in creating the future.

Let us start by looking at some very general features of synergy which we would expect to see manifesting in society. First, the essence of high synergy is that the goals of the individual are in harmony with the needs of the system as a whole. As a result there is minimal conflict between the elements of the system, and between these elements and the system as a whole. Evidence for such decreased conflict in human societies has already been found in studies of some tribal groups that have a naturally high level of synergy; very little aggression was apparent between individuals, or between individuals and the group. In a global high-synergy society we might similarly expect a large reduction in crime, violence, international hostilities and terrorism. Once we become aware that we are all of the same spirit, all human beings would become universally sacred; war, murder, mugging, rape and any other form of personal violence would become anathema.

Secondly, a high-synergy society would come about through a widespread shift from a derived sense of self to the universal Self. As a result we would begin to feel for the rest of the world in much the same way that we *feel* for our own bodies. This would almost certainly have a profound effect on the way we treat the environment.

At present, most of us have a spontaneous 'gut' reaction to the idea of deliberately harming our own bodies – chopping off a finger, for example. This is because we know, in a very intimate way, that the finger is part of us. If we begin to feel the same way towards the rest of the world, we will know, not just as an intellectual understanding but as an immediate inescapable awareness, that all aspects of the world are as much a part of ourselves as our bodies are. We would then find it as crazy to decimate the equatorial rain forests for some short-term end, as it would be to cut off a finger simply because it hurts or is in the way. In short, humanity would begin to live in harmony with itself and the environment.

Thirdly, we could also expect a reversal of the many inappropriate, wasteful, and often damaging behaviours, which we earlier saw to be the consequence of a derived sense of self. Being no longer dependent upon our interaction with the world for our sense of personal identity, we would no longer need to search for positive psychological support, and would not be emotionally hurt by negative criticisms. We would not need to gather excess possessions, to be seen to belong to the right group, or adhere to certain beliefs to prove who we are. No longer continually seeking to reaffirm our sense of self, we would be able to act more in accord with the overall needs of a situation rather than our ego's needs.

Furthermore, the widespread development of higher states of consciousness would result in a society where spiritual values were a universally accepted part of life. Self development and inner growth would be recognized as the proper end of all human striving, and the basis of our continued evolution.

However, being in tune with each other, humanity and the rest of the environment does not mean we would all become similar, either in behaviour or needs. The cells in your body do not have to become similar in order for you to be a healthy organism; the oneness is at a far deeper level. In a high synergy society, there would be just as rich a diversity of people and interests as there are now. Indeed, freed from the psychological need to belong and conform to a norm, people would be at greater liberty to express their individuality. Rather than everybody tending to become more alike, diversity would increase. It would be seen as a healthy and productive aspect of an evolving organic society.

Similarly, there would be no loss of diversity at the national level. If anything groups would become more in touch with their particular ethnic and cultural heritages. This trend towards increasing diversity in no way contradicts the complementary trend to greater collectivity, and the formation and expansion of trans-national groups such as the EEC. Just as in our own bodies the heart, lungs, kidneys and liver function with a high degree of autonomy, and simultaneously work together as part of a larger whole; so in a high synergy society there would be a synthesis of autonomy and co-operation at all levels from the individual, through the family, community, state and nation, to the whole world. Social synergy would not therefore imply any form of totalitarian world government.

Nor would a shift to high synergy would not mean that the multitude of problems now facing us will suddenly, and magically, disappear. Pollution, starvation, energy shortages, mineral shortages, unemployment, poverty, crime, and social, racial and sexual inequalities, must all be attended to. This will still require individual action, reform movements, and pressure groups. We would need to put as much, and probably more, effort into resolving these issues as we do now.

Without a major shift in consciousness, however, the solutions we may come up with will only succeed in building different structures within an already faulty framework. They will be incomplete, and the chances are that sooner or later they will fail, just as the present approaches are failing today.

Many of the long term goals in a high-synergy society may well be the same as those of contemporary society, e.g. improved health, better nutrition, more efficient use of energy and mineral resources, etc. The difference is that from an enlightened state of consciousness such goals would not just be intellectually understood as desirable: they would be positively desired.

NO LIMITS TO GROWTH

One very probable consequence of this shift would be a major change in society's attitude to growth. At present most people see growth in predominantly material terms. With a general shift towards higher states of consciousness we would begin to see growth in a much wider context; personal and spiritual growth would become as important, if not more important.

Reports such as *Limits to Growth* by 'The Club of Rome' (an international group of scientists concerned with the future welfare of humanity) have made clear that material growth cannot continue forever; quite soon we are going to start running out of many essential resources. Even before we run out there comes a point where continued growth will entail unacceptable social, political and environmental costs, and we may already have reached these operational limits to growth.

Some people feel that we must therefore deliberately curtail our urge for growth, and learn to be satisfied with what we have. This feeling has become manifest in the movement towards what is known as 'voluntary simplicity': thoughtful consumption, resistance to artificially created 'needs', and sensitivity to limited energy and mineral resources. Moreover such values are becoming increasingly widespread. Recent polls by Gallup, Harris, and others have found that about 70% of the US population would rather learn to appreciate more human, less materialistic, values than continue to accrue more goods; over 50% think that 'doing without something and living a more austere life would be a good thing'.

Voluntary simplicity is certainly much more attractive than forced simplicity – yet it still has feeling of restraint about it, a 'doing without something'. In a high-synergy system we would see another form of simplicity – spontaneous simplicity. This would not involve any form of voluntary restraint or resistance to perceived needs; instead it would represent a major shift in needs.

Abraham Maslow identified what he termed a hierarchy of needs. At the bottom of the hierarchy are the basic physiological needs for food, water and oxygen, which ensure our day-to-day survival. When these are satisfied, we turn to the second level, the need for warmth, safety, shelter, clothing and long-term survival. At the third level is the need for love and procreation, ensuring the survival of the species. Once this has been looked after, people look for esteem and social status. At the top of the hierarchy comes the need for self-actualisation and enlightenment.

Many of us in the developed countries have become stuck at the fourth level of need, the need for esteem and status, and we usually try to satisfy this by gathering wealth and material possessions. When we fail to feel fulfilled, we make the error of thinking that if only we had more or better possessions, everything would be fine. Yet, many of us already have enough in the way of material goods, and our continued attempts to satisfy, or rather to over-satisfy, the fourth level of need, result in very little real satisfaction. More often we feel empty and unfulfilled.

If, however, higher states of consciousness were to become the norm, the root cause of much of our over-consumption would have died away. No longer bound by the needs of the derived self, we would naturally stop consuming products and services we do not really need. We would realise that, by and large, we already have enough to satisfy most of our material needs, and would move on of our own accord to higher levels of satisfaction – self-actualisation. Society as a whole would step up the hierarchy to the fifth level of need.

It is not, therefore, the urge for growth itself which is at fault but our limited awareness of possible avenues of growth. The movement towards self-realisation would allow us to expand our concept of growth. No longer would we have to curb growth by repressing our material desires (just as when the first level, hunger, is satisfied most of us do not have to repress the desire to eat). Growth would have been raised a level rather than curtailed. On this basis material consumption could start to decrease without any loss of human fulfilment.

UNEMPLOYMENT REVALUED

The shift in needs from self-esteem to self-actualisation will bring many major changes in values. One change in particular will be in our attitudes to work. Traditional areas of employment are almost certain to decrease in the future. Increasing technological improvements and automation, in a diverse range of occupations, mean that society will not need everyone to work full-time. If, in addition, there is a significant shift towards higher states of consciousness, and a consequent decrease in our material needs, employment will drop still further.

The idea of unemployment still has many negative connotations. These mostly stem from times in the past when it may have been necessary for everyone to pull their weight, in order that there should be enough food and basic commodities. Although this is no longer true today, it has left us with the attitude that employment is 'good', and unemployment 'bad'. More often than not, the unemployed have been regarded as second-class citizens. Unemployment has, consequently, become as much a personal crisis as a financial one. A major source of status has been lost, and as if that were not bad enough, the person has also to cope with the stigma of being 'unemployed'.

Although some people may perceive a conflict between the right to work and decreasing job opportunities, the real conflict is between the need for status and decreasing opportunities to satisfy this need through employment. In a high-synergy society, the need to reaffirm ourselves through our social status would have died away, and this conflict would become considerably less dominant. Indeed, it would probably be considered strange to want to work when there is so much other work to be done on the self.

We could also expect to see a major re-evaluation of what constitutes employment. Many activities, which today are not paid because they are not in the short-term interest of any individual company or institution, would be appreciated for their value to society as a whole. A person may be benefitting society in the long term in any one of a number of ways not currently regarded as gainful employment – by furthering his own education, passing on his wisdom to others, contributing to the artistic and cultural heritage, or even sitting deep in meditation. As society came to value such activities, the current distinction between salaried employment and unemployment 'pay' would have to be reconsidered.

At present work is often used as a way of occupying time. Because most people's awareness is essentially outer-directed they need to fill their time with experience. Eight hours of work conveniently fills much of the day; while at home television employs time admirably, as do most other forms of entertainment. Hobbies, housework, social gatherings, conversation, and a few socially acceptable drugs fill in the rest. Clearly many people appear to go about their lives on the assumption that they should be left alone with their own self as little as possible.

As needs shifted up to the fifth level, to self-actualisation, we could expect to see a growing use of time for inner development. Rather than spending their spare time trying to lose themselves in the world of experience, many people would be glad of the opportunity to withdraw from the external world more often in order to explore their inner selves. There would be a shift from the 'right to work' to the 'right to be'.

Much of the increased time available might also be spent in education, and not just in the sense of learning and schooling but in its fuller sense of unfolding potential. Education would become a lifelong activity, rather than simply a preparation for adulthood.

Currently, most of our intellectual and mental abilities virtually stop growing around the time we finish our formal education – unless, that is, we are engaged in activities which require continual updating, re-education, challenge and stimulation. With lifelong education the opposite trend would occur: a continued growth and unfolding of our innate, and largely untapped, potentials, would become the norm rather than a privilege.

Moreover, the current emphasis of education on facts and information would give way to a balance between the development of knowledge and the development of the knower. Society would enter a New Renaissance as creativity, intuition and personal development became valued as highly as science, technology and economic development are today. Technological progress would not be seen as a threat to the quality of life, but as a liberator, allowing people to move on in the direction of self-actualisation, thereby improving the quality of life in the most fundamental way possible.

Earlier in our history, the division of labour and widespread industrialisation freed many people from the need to work on the land, allowing them to spend more time furthering their material growth. Today, the increasing application of technology and automation is freeing us from the need to perform tedious manual work, giving us the opportunity to move on to inner growth. In this respect, the reduced need for employment is in line with the basic thrust of human evolution – the inner evolution of consciousness.

HEALTHY, HOLY AND WHOLE

We ought also to expect a high synergy society to be a healthy one – the concepts of synergy and health are, as we saw earlier, intimately related. At present 'health' is usually used to mean an absence of any symptoms of disease or sickness. Providing your body is functioning reasonably well, temperature, pulse and blood pressure normal, no recurrent pains, rashes or fainting spells, you are well.

True health, however, means much more than this. The root of the word 'health' is the Greek *holos*, meaning whole, and this is also the original meaning of the Anglo-Saxon word 'well'. Moreover, the word 'holy' comes from the same root. The healthy or well person should, therefore, be a whole person – one fully developed and integrated in mind, body and spirit. To heal should be to make whole; and the whole person is a holy person – spiritually mature, that is enlightened.

In a spiritually transformed society this relationship between synergy and improved health would probably manifest in a number of ways. First, most techniques which lead to an experience of the pure Self involve a physical relaxation and quietening of the mind. One general conclusion of the considerable research conducted on meditation, yoga and similar techniques is that they produce the exact opposite of the stress response. Blood pressure, heart rate, muscle tension and other variables commonly associated with stress decrease, as do also the levels of various 'stress hormones' in the blood. It is now widely recognised that stress is implicated to some extent or other in the majority of illnesses, both physical and mental. People practising such techniques should therefore be not only more relaxed, but less prone in general to illness – and the few studies which have been conducted in this area tend to support this hypothesis.

Secondly, in addition to decreasing physiological stress, the movement towards higher states of consciousness would also mean that fewer situations were perceived as stressful. This would come on the one hand through the decreased conflict and aggression characteristic of high synergy; and on the other hand from a marked reduction in psychological threats. The vast majority of such threats are only threats to the derived sense of identity. Once the identity has shifted to the pure Self, a major factor in stress would have been eliminated.

Thirdly, there would be fewer man-made health problems. Some of our present ill-health is attributable to the exploitative aspects of a low synergy society: pollution in the atmosphere; toxic wastes finding their way into the drinking water; cancer-inducing additives in foodstuffs and various commercial products; cigarettes, alcohol, sweets and other known health dangers promoted for the financial benefit of a select few; etc. These are all factors which would decrease in a society living more in harmony with itself and the world around.

Fourthly, we could expect a shift in medical care towards holistic health practices. Western medicine is largely based on the 'Humanity versus Nature' approach, as is evidenced in the widescale use of antibiotics, surgery and radiation treatments. Such approaches have met with significant successes, but they also produce unwanted side effects. (It is estimated that one third of all hospital admissions in the USA are due to iatrogenic disease: that is, disease caused by previous medical treatments.) A common experience of many people engaged in various forms of inner development is that they become more aware not only of their unity with the rest of the world, but also of the interplay of mind and body. It starts to become very clear that treating only the physical symptoms, and ignoring their psychological and spiritual correlates, is not treating the whole system.

One effect of this growing awareness is a greater respect for cne's own body. At present many of us exploit our own bodies in our search for ego-affirmation, eating the 'right' (but often wrong) foods, inviting skin cancer in order to look bronzed, abusing ourselves in our efforts to gain attention, or temporarily escaping from the misery of a limited identity in a stiff drink. With the derived self no longer dominating activity, such behaviours should become far less prevalent. We would care for ourselves more. Such caring is the real basis of preventive medicine and the essence of holistic health. It is, in the words of the philosopher Henryk Skolimowski, 'taking responsibility for the fragment of the Universe which is closest to one, expressing a reverence towards life through oneself'.

In addition, holistic health gives greater recognition to the largely unfathomed healing potential of the mind itself. The little research that has been done in this area suggests that we may all have the ability to heal ourselves of anything from a common cold to cancer (and we shall be looking a little more into this in the next chapter). Furthermore, the attitudes of mind which most help this ability are very similar to those found in meditation – a state of relaxed attentiveness.

LEFT *AND* RIGHT

Being healthy and whole applies as much to the brain as to the rest of the body, and this is another area where we might well expect functioning to become more integrated. Since the mid-1960s a variety of psychological studies have shown that the left and right sides of our brains specialize in different types of activity. The left side of the brain appears to be more concerned than the right with rational, sequential thought, and with linguistic faculties such as reading, writing and speech. The right side of the brain seems to be more concerned with visual-spatial functions, aesthetic and emotional appreciation, and perhaps with intuitive thought. The general picture which has emerged is that the left is more analytic, processing in a step-by-step mode, while the right is more synthetic, processing more holistically. In additon, the left side has been associated with active modes of thought, with 'doing'; and the right with receptive modes of thought, with 'letting things be'.

In most modern societies, people tend to use the functions associated with the left side of the brain more than those of the right. This is reflected in our general approach to the world, the activities and professions in which we engage, and the types of mental activities we value and encourage. When, for example, we say that someone has 'a good mind' we usually mean to imply that they can think logically, reason well, and express themselves clearly – all predominantly left-brain activities.

This preference for the left is partly a reflection of our educational systems. Most people have been taught how to use and develop the functions associated with left brain (i.e. the 3 Rs – reading, writing and 'rithmetic) more than those of the right. It is also partly cultural. Our emphasis on doing and achieving, rather than being, has reinforced left-brained ways of thinking.

Studies of the electrical brain activity of people in deep meditation have revealed a progressive synchronisation of the electrical activity coming from either side of the brain; and the deeper the meditation (subjectively speaking), the greater the integration. This synchrony suggests an increased balance between the two modes of thinking. It would seem likely, then, that in the enlightened state, thinking would be both analytic *and* holistic, intellectual *and* intuitive, active *and* receptive.

In many cultures the faculties associated with the left and right sides of the brain parallel male-female symbolism. Masculinity is associated with active, doing, intellectual modes of operation; femininity more with passive, receptive, intuitive modes.

Taking humanity as a whole, the dominance of the left-brained approach suggests that the global brain is also left-dominant, and this becomes manifest in the masculine nature of most contemporary societies. Most nations seem preoccupied with science, technology, rational thinking, and with making things happen. The feminine principle, on the other hand, symbolises the energy of the living Earth, the creation and sustenance of life, the ecological principle, living in harmony with the planet. This is the currently under-used 'right side' of the global brain. In this respect the rising wave of feminism may be more than an overdue revolt against a male-dominated society: it may also be a sign of a growing shift of consciousness, both in the individual and society.

With a widespread shift in consciousness, and the integration of the male and female within us all, we might expect social attitudes and values to become androgynous. This does not mean neutral, but an integration of the qualities of male and female. The important differences between masculine and feminine ways of thinking would be recognised and given value. There would be a balance and synthesis of the left and right sides of the global brain.

SYNCHRONICITY RULES

Another major consequence of a widespread shift in consciousness could be an increase in synchronicity; an increase in curious and inexplicable chains of coincidences. Indeed, synchronicity would probably become the norm. Such a development may be much harder to accept than some of the others we have already discussed, but it would be a natural correlate of humanity moving towards a social superorganism.

The celebrated Swiss psychologist Carl Jung described synchronicity as an acausal connecting principle at work in the Universe. Acausal events are ones with no apparent physical connection, and no way in which they might affect each other. But such events may sometimes be connected in a way which is very meaningful and significant for the people concerned. This is synchronicity – a *meaningful* coincidence. As such it is different from synchronism, which is merely the simultaneous occurrence of two unrelated events.

As an example of synchronicity Jung quotes the following case:

A certain Monsieur Deschamps, when a boy in Orleans, was once given a piece of plum-pudding by a Monsieur de Fortgibu. Ten years later he discovered another plum-pudding in a Paris restuarant, and asked if he could have a piece. It turned out, however, that the plum-pudding was already ordered – by Monsieur de

Fortgibu. Many years afterwards Monsieur Deschamps
was invited to partake of a plum-pudding as a special rarity.
While he was eating it he remarked that the only thing
lacking was Monsieur de Fortgibu; at that moment the door
opened, and an old man in the last stages of disorientation
walked in; Monsieur de Fortgibu who had got hold of the
wrong address and burst in on the party by mistake.

Strange as they might sound, such coincidences are not
uncommon. Alan Vaughan, in his book *Incredible Coincidence*,
details many such instances. He tells, for example, of a lady
who had locked herself out of her house, and was busy trying
to find another way in, when the postman arrived with a letter
from her brother returning a spare key he had borrowed.
Another typical case is of the person who accidentally got off
the New York subway at the wrong station, realised his
mistake when he reached the exit, and was about to return to
the trains when he bumped into the very person he was on his
way to visit.

Now it could be argued that such coincidences are not of
any statistical significance. We should expect they would
happen once in a while – for each time a person has a
fortuitous meeting, there may be a hundred or a thousand
times when no coincidences take place. It is, however,
practically impossible to gather all the statistics necessary to
evaluate this line of argument. Estimating the likelihood of a
particular strange coincidence is not very difficult; Vaughan
himself carries out such an analysis on several cases, and finds
the odds sometimes of the order of a trillion to one against.
The difficulty lies in assessing all the possible, though
unexpected, coincidences which could have occurred but did
not.

There are, however, two general characteristics of such
experiences which do very much suggest they are more than
chance, and which also have important implications for a high
synergy society.

First, the outcome of such a coincidence generally appears
to be beneficial to the person involved, fulfilling his desires or
needs at that time. Moreover, one does not benefit from a
coincidence at the expense of other people; usually everyone

involved finds their own particular needs being fulfilled by the interaction. If they were only chance occurrences, we should expect as many negative as positive outcomes. Yet this does not seem to be the case. Some negative instances have been reported, but they are not so common; and it seems unlikely that people only notice the positive ones. In most cases it appears that the environment is acting in a very supportive manner. Far from being random events they appear to have a benevolent nature.

Secondly, it seems as if the rate of occurrence of such coincidences can be directly influenced by the state of mind of the person involved. This is not to say they can be deliberately willed to happen. Indeed deliberate willing could be counter-productive. The man who got off the subway at the wrong station would almost certainly not have done so if he had been consciously trying for such a meeting. Trying is doing, a form of activity in which the individual is manipulating the world. This is the wrong state for such experiences, which seem to occur more frequently when one is in a receptive state, open to some form of unconscious decision making, and flowing *with* the world rather than pitting oneself against it. The occurrence of synchronicity can therefore be encouraged by a relaxed, peaceful state of mind – one which is similar to the state of mind brought on by meditation.

Many people who practise meditation of one kind or another have found that the deeper and clearer their meditations, the more they experience curious patterns of coincidences. This tends to be particularly so after extended meditation retreats; on returning to regular activity, each day can seem like a continual train of the most unlikely, and most supportive, coincidences.

Sceptics might argue that people in these circumstances are just more open to noticing synchronicity; but when the coincidences are so remarkable and significant that they influence major aspects of one's life, it is difficult to believe they would otherwise pass unnoticed.

This relationship between synchronicity and one's state of mind is not a new finding. Two and a half thousand years ago the Upanishads of ancient India observed that:

'When the mind rests steady and pure, then whatever you desire, those desires are fulfilled.'

Similarly in Christianity we find the twentieth-century British archbishop William Temple reporting that; 'When I pray, coincidences start to happen. When I don't pray, they don't happen.' This observation suggests that it may not be the particular supplications in a prayer which are important but the state of consciousness which is produced. Many religious teachings hold that 'true prayer' is not so much to beseech God for favours, but to quiet the mind and open it up to deeper levels of consciousness.

We might expect, therefore, that as more people begin to raise their level consciousness, synchronicity would become a much more widespread occurrence. Indeed, some people would claim this is already beginning to happen. For instance, at Findhorn – a community of several hundred people in the north of Scotland, which focuses on inner growth and loving work – such chains of coincidences are accepted as part of life. Some people would argue that we should start to worry when they do not happen, for that is a sign of lack of inner attunement. Thus if higher states of consciousness were to become a reality, we could envisage a society in which supportive coincidences were no longer marvelled at, but accepted as the natural order; a society in which things tended to work out better for everybody.

This trend is also predicted by a consideration of a high synergy society as a healthy social super-organism. Returning briefly to the case of a cell in your body, let us consider how it might, if it were aware, likewise experience a form of synchronicity. It might notice that the blood always seemed to supply the oxygen and nutrition it needed, when it needed them, simultaneously removing waste products as they built up. Such a cell might well marvel at the incredible chain of coincidences which kept it alive and provided spontaneous support to most of its desires. Everything would probably seem to work out just right. Its prayers would be continually answered. It might even suppose the existence of some individual answering agency or god.

We, however, looking at the situation from the higher

perspective of the whole organism know that what the cell perceives as a chain of 'curious coincidences' could in fact be ascribed to the high synergy which comes from the whole body functioning as a single living system. The cell may not be directly 'aware' of the body as a living being, but it would nevertheless benefit from the high synergy which results from this wholeness. Furthermore, the healthier the body is, the more supportive coincidences the cell would notice.

What we regard as curious chains of coincidences may likewise be the manifestation at the level of the individual of a higher organising principle at the collective level – the as yet rudimentary social super-organism. As humanity becomes more integrated, functioning more and more as a healthy high synergy system, we might expect to see a steady increase in the number of supportive coincidences. A growing experience of synchronicity throughout the population could, therefore, be the first major indication of the emergence of a global level of organisation.

ESP AND THE MIRACULOUS

Another aspect of a shift towards high synergy might be an increase in paranormal phenomena such as telepathy, clair-voyance, and precognition – what are collectively known as extra-sensory perception, or ESP. Despite numerous attempts to find a cause-effect explanation for such phenomena, no one has yet satisfactorily explained how they might occur. To some this might seem sufficient reason to reject their validity; but others such as Jung have seen them to be examples of synchronistic phenomena – an indication of a higher organis-ing principle beyond the causal space-time arena with which most science deals. If so, an increase in ESP would be another aspect of a general increase in synchronicity.

To some people ESP may suggest the idea of being able to 'read another person's mind', or 'predict the winner of a horse race'. While such things may be possible, this is not the way such phenomena usually manifest. Telepathy, for example, means literally 'feeling at a distance' (*tele-pathos*); and it does indeed seem to happen more often on the feeling level – one is more likely to have a feeling that a close friend is ill, rather than a sudden clear message from the friend.

It might be thought that such abilities, if they do exist, are the attributes of very few. But recent research is suggesting that we may all have these faculties latent within us. Targ and Puthoff at the Stanford Research Institute, for example, have been investigating what they call 'remote viewing'. This is the ability to describe the scene at an unknown, randomly chosen location. They began the experiments with people who already seemed to have well-developed psychic abilities, but later found that anyone could be equally successful in describing the target location. The office secretary, for example, who claimed no special abilities, scored as well as the acknowledged psychic. What seemed to be important was a general willingness and openness to explore in greater depth some of the faint images and hunches which often pop up in the mind, and which we ordinarily might reject as spurious or irrelevant. Studies such as these suggest that ESP almost certainly occurs much more widely and frequently than we are aware of.

Other experimental work suggests that it is the more visual, right side of the brain, rather than the verbal-analytical, left side which is involved in these phenomena, and that it may have been our tendency to concentrate more on the skills associated with the left hemisphere that has resulted in many of us not experiencing telepathy and other forms of ESP. First, people are generally much more accurate in ESP experiments when asked to describe the images they have, rather than their verbal ideas. Secondly, people with left-brain damage are often better in ESP tasks, probably because there is then less interference with the right-brain functions. And thirdly, a receptive state of mind (a characteristic of the right-brain) appears to be the most conducive to ESP. As with other forms of synchronicity, it is very difficult to make these things happen.

If, therefore, higher states of consciousness do lead to an integration of the left and right sides of the brain we might also expect them to lead to a more widespread occurrence of ESP. This contention is further supported by the claim of most spiritual teachings that these abilities will develop quite naturally as one's level of consciousness rises.

ESP is not, however, the only paranormal faculty likely to develop in a society of spiritually enlightened souls. The writings of most spiritual traditions and many mystics suggest that various other abilities would emerge, some of which might make ESP look like spiritual kindergarten.

Indian traditions, for example, speak of powers called 'siddhis', which come as a result of enlightenment. The *Yoga Sutras*, an ancient text which is the cornerstone of yogic philosophy, describes some fifty-two such powers, ranging from telepathy and clairvoyance, to invisibility, levitation, walking on water, and being in two places at once.

The *Anguttara Nikaya*, a collection of the Buddha's teachings, describes similar supernormal abilities: 'There is the one who . . . having been one becomes many . . . appears and vanishes, unhindered he goes through walls . . . He dives in and out of the earth as if it were water. Without sinking he walks on water as if on earth. Seated cross-legged he travels through the sky like a winged bird'. Not only was the Buddha himself said to be endowed with such powers, but so were hundreds of his monks.

In Christian scriptures we find Christ displaying similar powers. This is sometimes seen as proof of Christ's divinity, yet he himself claimed that such capacities were open to anyone – 'You shall be able to do all these things and more'. Peter, once he had seen Christ walk on water was able to do it himself – until, that was, he 'lost faith'. Many holy men and saints were said to have performed similar miracles – St. Theresa of Avila and St. John of the Cross, two medieval Christian saints, were both prone to levitation. Indeed, so much are such powers seen to be a natural outcome of true spiritual development, the Roman Catholic Church has made the perfomance of miracles a prerequisite for official canonisation.

Physical science remains unable to explain how such phenomena could possibly take place – in many instances they directly contradict the current paradigm. Yet so many teachings from around the world affirm their occurence, that it would be foolish to dismiss them completely, even if we cannot yet understand them, or experience them ourselves.

The implication of such claims is that a society of enlightened people could be a society in which we all had such faculties. Unbelievable? Impossible? Or are they an indication of just how profound the transformation could be – a water molecule's glimpse of steam?

Whether or not such abilities would be part of a spiritually transformed society, and whatever other as yet unforseen developments might occur, we are still left with the question: Could it really happen?

Granted, there are a number of indications that humanity is heading towards a major evolutionary leap, and we have seen that such a transition, if it were to occur, might not be many years away. But it certainly is not inevitable.

Yet neither is the future fixed. We shall see that the choice of whether or not we do move in this direction rests very much with us.

CHAPTER 13

Choosing the future

Shortage of time is the greatest shortage of our time.

Fred Polak

Throughout this book I have been taking a very optimistic view of humanity and its future – and deliberately so. Why? More than because optimism is enjoyable, and more than because I very much hope for this kind of future. I have taken a positive perspective because I believe that the image we hold of the future plays a role in helping that future to emerge. If I entertained a negative scenario, and encouraged others to hold that view, I would be instilling a negative mental set about the future, and would be helping to make that future more likely. By the same token, encouraging positive visions of the future may actually help us move in a more positive direction.

Our dominant image of the future today is generally a pessimistic and depressing one. The majority of people take it for granted that the chances of some form of collective calamity are high. Moreover, most of the information we receive about the world supports this negative image. Newspapers and news bulletins effectively chronicle the problems and disasters of the day. News, it seems, is bad news.

As society becomes increasingly despondent about its own future, it seems to produce more stories of doom to reinforce its gloomy mind set. (For example, a recent survey of the films being shown in London (excluding the pornographic films) revealed that over 80% involved disaster, disruption or violence of one form or another.) The net result is a reinforcement of the image of society moving in the direction of collapse. Negativity breeds negativity.

Indian teachings have summed this up in the saying 'Sarvam-annam' – 'All is food'. By 'all' is meant not only the food we eat and the air we breathe, but what we take in through the senses as well. Negative, destructive or aggressive experiences are going to have a negative, destructive or aggressive affect on our consciousness. In effect we pollute our minds as much as we pollute the physical world around us, and this mental pollution can affect our lives as radically as physical pollution, if not more so.

A study in Great Britain showed that the violence of news reports led to a more violent attitude in children who viewed them. Those who watched the local Northern Irish news bulletins, which contained about four times as many references to violence as the BBC national news, developed a considerably more violent attitude than those who watched the national news, irrespective of whether they lived in Northern Ireland or mainland Britain.

Yet, contrary to what many news editors apparently believe, a study in the USA found that when negative news bulletins were replaced by positive ones, people found the news just as enjoyable. More significantly, they began to change their attitude towards the people they met in daily life, seeing them in a much more positive light.

But the sets we have of society can have an impact that goes far beyond our individual attitudes and behaviour. In his book *The Image of the Future* Fred Polak, a Dutch futurist, showed that our images of the world play a crucial role in shaping society. He found that in every instance of a flowering culture there had been a positive image of the future at work. As an example, he noted the way in which the Jewish people have remained spiritually intact over centuries of adversity. Israel's power, he suggested, rested in her living image of the future. The power of the prophets and the revolutionaries came from a burning expectation for the future. When the opposite happened, when the images of the future were weak, the culture decayed – as was the case with the fall of the Roman Empire.

Polak also found that the potential strength of a society was reflected in the intensity and energy of its images of the future. These images acted as a barometer, indicating the potential rise or fall of a culture. This intimate relationship between the image of the future and the future itself made it possible to predict the direction cultures would take. He concluded that 'bold visionary thinking is in itself the prerequisite for effective social change'.

IMAGES THAT HEAL

The images we hold in our minds can also have a profound impact on human physiology, and may play a crucial role in healing. Much recent work in this area has centred on the role of mental images in the treatment of cancer, and it will be worth considering this research briefly, since it offers clues as to another possible approach for healing the planetary cancer.

Western society's general attitude towards cancer is a negative one: it is seen as a potentially fatal illness, widespread and difficult to cure. So negative are our sets that in some areas of society it is taboo to mention the subject, or to admit that one personally has some form of malignancy. For the cancer patient this negative image is compounded by prognoses about his chances for survival. If he is given the mental set that he has six months left to live, and this is backed up with the authority of doctors, he usually fulfils the predictions – a fact that aboriginal witch-doctors know only too well.

At the same time, however, there have been a number of well-established instances of spontaneous remission, in which the cancer patient gets better of his own accord in spite of a negative prognosis. In many of these often dramatic recoveries it was found that for one reason or another (very often as the result of a deep spiritual experience), the person changed his whole attitude towards life and regained a strong will to live. Generally speaking, the more positive his outlook the greater the chances of remission.

This healing power of imagery has been taken further by several cancer specialists, who are finding that many malignancies seem to respond positively to imagery. Two of the pioneers in this field were Carl and Stephanie Simonton, working in Fort Worth, Texas. They found that if a patient is taken into a state of deep relaxation, and in this state visualises his white blood cells swarming over the cancer cells, consuming them and carrying them away, then very often the malignancy ceases growing and starts shrinking. In many cases it has disappeared completely.

They also noticed that the more hopeful the imagery, the better the result. If, for example, a patient visualized the cancer as a large logjam blocking a river, being attacked by just one man – a single white blood cell – then the imagery conveyed little hope of success and was not very effective. But, if the white blood cells were seen as a vast army of white knights on horseback charging through the landscape killing the much smaller, slow-moving cancer cells, the imagery had a much more powerful effect. Optimism, it appeared, is crucial.

In many respects humanity itself is behaving rather like a malignant growth on the planet, and it was suggested earlier that both the general behaviour and the root causes of humanity's malignant tendencies bear close similarities to cancer in the body. Perhaps then, some of the principles of the Simontons' work may be applicable to helping heal the planetary cancer.

There appear to be two key elements in their approach: (1) a relaxed, almost meditative, state of mind; and (2) a specific image of the desired result. One of the effects of relaxation is to increase the synchrony of the brain activity. The firing of the billions of individual cells become more in step with each other, giving rise to strong regular rhythms of electrical activity. What would produce a corresponding synchrony in the global nervous system? From our early discussions, the answer could be people meditating together, coming into tune through the universal Self.

What then might be the effect of a million people across the world, meditating simultaneously, and from the depth of their meditation visualising, for example, the whales of the world no longer being harassed, but growing in numbers and cared for by humanity; or maybe visualising the long-term peaceful settlement of some international dispute, seeing the parties concerned agreeing on a mutually satisfactory solution? Might we see a collective synchronicity? Might there come an end to whaling or the settlement of the dispute?

The answer is not at present clear. Some who have tried such collective imagery do claim positive results, but, with a general lack of well-controlled studies, sceptics could justifiably claim that any apparent results might only be coincidental. One could argue, however, that we might expect only marginally significant results at present. First, there is still the collective inertia of the 'old consciousness' to be battled against. Secondly, the experiments so far have not generally involved very large numbers of people, nor have the images always been that specific. And thirdly, the techniques of meditation employed may not always have produced sufficient opening to the more universal levels of consciousness.

Thinking of a tumour dissolving is not as powerful as sitting in a particularly quiet state with specific images of white blood cells battling and eliminating the malignant cells. Similarly, getting people simply to sit down at a certain time and imagine world peace, valuable though this may be, is not likely to be as beneficial as using a specific techniques to produce the right state of consciousness, and then using a very specific mental image. Hopefully, however, as psychotechnology becomes an established and productive area of scientific research, we may discover just what states of consciousness are best for this sort of endeavour, how to induce them most effectively, and the most beneficial types of imagery.

This proposal may sound 'way out' to some, and only time will tell if it is possible. But if such approaches *are* found to have significant effects they could become a very powerful adjunct to meditation itself in furthering planetary healing. Indeed, they could be one of the most powerful agents for change that humanity has ever had at its disposal.

OUR EVOLUTIONARY TEST

To hold a positive image of the future does not mean that we should fill our minds with naive optimism, and sit back hoping that all will be well. Humanity is indeed in a state of severe crisis, and there is no law of nature that says we will necessarily survive. Even if humanity does experience the kind of transformation envisioned here the current problems are almost certainly going to get steadily worse, and it may well be that we have to descend into some very major global instabilities before a new level of integration finally emerges.

From the perspective of dissipative systems theory these crises can be seen as the catalysts pushing humanity on to a new evolutionary level. If humanity successfully adapts to the crises it may break through to a higher level of organisation. But if it fails to adapt it may, if the crises are severe enough, break down and collapse completely.

There are indeed any number of collective disasters that could befall us before the general level of consciousness has risen sufficiently to bring about the needed transformation. Even if we do manage to avoid these catastrophes, many other negative scenarios are still possible: increasing terrorism, crime and personal violence; nations fighting each other as they greedily grab what they can of dwindling resources (a fight that has already begun), economic collapse bringing roving hordes out of the cities in search of food; ghettoes spreading across continents. Alternatively we just may manage to continue on our present path, dealing with today's problems in the same partially successful ways we have tried so far. Humanity might not collapse, but neither would it move on to become an integrated social superorganism.

If we do not make the transition it might be thousands of years before humanity stands upon the threshold again. Or it might never happen with the human species. If we wipe ourselves out it could take millions of years for another species to evolve with the same potentials. It might never happen on this planet. But there are plenty of other planets in our galaxy, and plenty of other galaxies. The Universe will carry on evolving towards higher levels of integration and complexity whether we do or not.

If, however, humanity does find ways to resolve the various problems and conflicts facing it, it will have proved it can adapt successfully. In this respect crises serve not only as evolutionary catalysts, but also as evolutionary tests, examining the adaptability and viability of the system. Humanity's currently growing set of crises could well be seen in this light: we may have reached the final test of our viability for further evolution.

This test is not a physical test; it is a test of our consciousness. It is a test of whether or not humanity is psychologically and spiritually fit to live on planet Earth; a test of whether we can change at a very fundamental level the way we relate to others and the environment; whether we can work in harmony rather than conflict; whether we can balance centuries of material progress with an equal inner growth; whether we can connect with that level of unity that we know theoretically (and, in those privileged, magical moments, know experientially) lies at our core.

Moreover, this test has a time limit. We do not have aeons to experiment; it is we who are alive today who must answer these questions.

Whether or not we pass is up to us. If we do pass, we may move into our next evolutionary phase – our integration into a single being. If we fail, we will probably be discarded as an evolutionary blind alley, an experiment which for one reason or another did not quite work out. Humanity will be spontaneously aborted, regardless of how close to the transition we might seem to be. If so, we will not be the first species to have become extinct on account of its failure to adapt.

Mother Nature, from her cosmic perspective, is not going to be too perturbed if we do not make it. She is not brought to despair by every blade of grass that is crushed underfoot, by every cell that dies, or by every seed that fails to germinate. Indeed, if humanity is aborted it will be for good reason. As far as Gaia as a whole is concerned, it will be as satisfactory an outcome as if we passed the test.

But the task of showing whether or not humanity is viable rests with us – each of us. Unlike other species, humanity can anticipate the future, make conscious choices and deliberately change its own destiny. For the first time in the whole history of evolution, responsibility for the continued unfolding of evolution has been placed upon the evolutionary material itself. We are no longer passive witnesses to the process, but can actively shape the future. Whether we like it or not, we are now the custodians of the evolutionary process on Earth. Within our own hands – or rather, within our own minds – lies the evolutionary future of this planet.

We can *choose* to carry ourselves through.

Will we be able to choose in time?

No one can say. But so long as the door is open to us, and so long as the evolutionary impulse shines through us, let us follow that inner urge. This is the cosmic imperative.

EPILOGUE

CHAPTER 14

Beyond Gaia

A walker in mountainous country, lost in mist, and groping from rock to rock, may come suddenly out of the cloud to find himself on the very brink of a precipice. Below he sees valleys and hills, plains, rivers, and intricate cities, the sea with all its islands, and overhead the sun. So, I, in my supreme moment of my cosmical experience, emerged from the mist of my finitude to be confronted by cosmos upon cosmos, and by the light itself that not only illumines but gives life to all.

From *Star Maker* by Olaf Stapledon

If humanity were to evolve into a healthy, integrated social super-organism, it would signal the maturation and awakening of the global nervous system. Gaia might then achieve her own equivalent of self-reflective consciousness, and a fifth level of evolution – the Gaiafield – might emerge. Gaia would become a conscious, thinking, perceiving being, and also a being functioning at a new evolutionary level with faculties quite literally beyond our imagination.

What will she discover, as she awakens?

To begin with she will start to become aware of her immediate environment – our solar system. She will begin to study the space around, the nourishing Sun, the other planets and their moons, looking to see if there are any signs of life out there. Indeed, this she has already begun.

Over the last two decades Gaia's nervous system has begun to sense the space around. A few thousand artificial satellites have been sent up; a hundred men have been into space, some of them to the Moon; probes have been sent to take close looks at Mars, Venus, Saturn and Jupiter, some looking for life; and other missions are planned for the sun and for comets. Seen

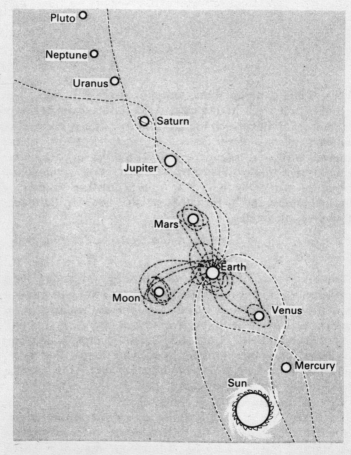

FIGURE 15. Gaia's growing nervous system; sensing her immediate environment through humanity's exploration of space. (Not to scale.)

from space it would look as if the Earth were beginning to grow nerves out into the solar system, fine tendrils sensing her immediate environment.

Already she has discovered that this solar system contains far more than the sun and its nine known planets. There are at least thirty-eight natural moons circling the planets, and thousands of asteroids in between the planets. In addition there are estimated to be several billion comets in orbit around the sun, some of which have orbits so huge they effectively extend the solar system halfway to the next star.

In addition to the many comets, moons and asteroids, there is the 'solar wind' – a stream of charged particles emitted by the sun – which flows far into the reaches of outer space. Also flowing out deep into space are the VHF radio and TV signals created by humanity over the last thirty years. Travelling at the speed of light Gaia's first emissions have already swept past the four hundred or so stars nearest to us.

If we could directly perceive the orbits of the comets, the solar wind, and the ever-expanding waves of radio signals, we would see our solar system, not as an isolated group of objects moving round the sun, but as a huge complex sphere of influence penetrating deep into the heart of other solar systems.

FROM GAIA TO GALAXY

Her explorations so far indicate there is little life elsewhere in this solar system; certainly not the rich biosystems from which other Gaias might emerge. But what of other solar systems? As Gaia continues her explorations, and looks beyond this solar system, will she find other planetary 'beings' out there, other evolved conscious entities also seeking contact? The answer could very well be 'Yes'.

Our solar system is minute compared to the whole galaxy. If we imagine the North American continent to represent our galaxy, then the Earth would be a mere ten thousandth of an inch, and its orbit the size of a pinhead; the sun would be the minutest speck visible to the naked eye in the centre of this pinhead; and the volume occupied by the whole solar system would be about the size of an apple – an apple hidden somewhere in North America.

According to current estimates, there are some hundred billion (10^{11}) stars in our own galaxy, a good percentage of which probably have planets accompanying them. Astronomical observations of the seventeen stars nearest to the Sun have found that at least four show signs of having planetary accompaniments. Furthermore, computer simulations of star formation suggest that the cloud of gas that surrounds a newborn star is very likely to condense into a planetary system, and the systems that result will usually be reasonably similar to our own solar system in their general structure – rocky, Earth-like planets near the star, and larger, icy, Jupiter-like planets farther away. Of these solar systems, it is estimated that in our galaxy alone there may be as many as 10^{10} with planets capable of supporting life as we know it.

How many of these planets would actually develop life? Maybe most of them. As far as the Earth was concerned, once the conditions for the emergence of life were right, life appears to have dawned very rapidly – it seems to have been a virtual inevitability. Furthermore, the Gaian tendency to preserve the optimal conditions for the maintenance and further evolution of life suggests that, once started, life was unlikely to die out. If these are general tendencies to be found throughout the Universe, we should expect life to emerge and evolve on virtually every viable planet, protected and nurtured by the planet's own respective 'Gaia'. Thus the number of potential Gaias within our galaxy could well be of the order of ten billion.

As we explored earlier, ten billion seems to be the approximate number of units required in a system before a new level of evolution can emerge. Could the possibility of ten billion living planets in our galaxy herald the emergence of some galactic super-organism whose cells are awakened Gaias?

Applying the same criteria as we used for the emergence of a social super-organism, we can see that ten billion Gaias distributed through a galaxy would not, on their own, create a galactic superorganism. There would also need to be a widespread communication and connectivity between the many Gaias, similar to the degree of complex interaction and organisation found in the human brain.

How might these Gaias communicate and interact? Inter-planetary expeditions would be far too slow – a single voyage across the galaxy would take millions of years. Electromagnetic communication, whether by light, radio, infra-red or X-rays would be much faster. Yet even at the speed of light it would take thousands of years for a message to cross the galaxy. This may only be a minute in the life of Gaia, but it is probably still too slow for a highly complex web of communication to emerge. Perhaps various forms of ESP are not limited by the speed of light; if so, they could enable much faster and more complex connections to develop. In addition, there could be various means of interaction characteristic of the Gaiafield itself that we cannot even conceive of, and these might well further enhance inter-Gaian contact.

If, in one way or another, Gaias were able to reach out, make contact and interact with each other, there could then come a time, millions of years from now, when inter-Gaian interaction and communication would have reached a suf-ficient degree of complexity and synergy for the ten billion Gaias in this galaxy to integrate into a single system. Our own solar system might no longer exist then; stars in the galaxy come and go as do the cells in a living organism. Even if our Gaia were still alive, humanity might have evolved beyond recognition, or perhaps new life-forms would have arisen, taking over humanity's role.

Regardless of when this might occur, this next evolutionary step would signify the transition to a galactic super-organism. The Galaxy would become her equivalent of conscious. With this would come the emergence of a sixth level of evolution; one as different from the Gaiafield as the Gaiafield is from consciousness, consciousness from life, and life from matter.

FULL CIRCLE

We likened the size of our own solar system in the galaxy to an apple in North America. Yet this galaxy is but itself a minute structure in the whole Universe – another apple lost in another huge continent. So huge are the dimensions involved that it is almost impossible to conceive of just how big the Universe is, and just what a minute speck in it our own galaxy is.

On a clear night we may look up at the thousands of stars filling the sky, yet with only one or two exceptions, every point of light we see, however faint, is a star within our own galaxy. We are seeing less than a *billionth* of the Universe. When we look into space through a powerful telescope, we find that the dark patches between those stars are filled with myriads of tiny points of light. Moreover, each of these specks is not a star but a whole galaxy. And these are just the galaxies close enough to be seen.

Looking at the distribution of these galaxies, astronomers have found that they are not scattered randomly throughout space; they tend to group together in clusters. Some clusters are small, containing ten or twenty galaxies, while others may contain as many as a thousand galaxies.

Our own galaxy is part of a small cluster containing twenty-seven known members, and all around our local cluster are similar clusters of galaxies. In the middle of these is a very much larger cluster, called the Virgo cluster, containing thousands of galaxies. Seen from far out in space, our local cluster appears to be part of a huge cloud of clusters, all centred on the Virgo cluster. This entire system is called the Virgo Super-Cluster. Looking out even farther into space, astronomers find the Universe filled with numerous other similar super-clusters, each containing thousands upon thousands of galaxies.

If we liken an entire galaxy to a single atom, then what astronomers are observing is reminiscent of the way in which atoms collect together to form simple molecules, which in turn group to form complex macromolecules. If thousands of macromolecules can build up a living cell, could the numerous super-clusters themselves be integrated into a single system? Could the Universe as a whole become a living system?

When, at the start of our journey, we looked at the possibility of our planet being a living system, we found a number of strange 'coincidences' which happened to be the optimum for the evolution of life. They were either a very unlikely, and very fortunate, series of flukes; or the planet was somehow purposefully maintaining this optimum. Physicists are now discovering that some similar strange coincidences exist in the Universe as a whole.

For some unknown reason, slightly more electrons than positrons (anti-electrons) were created in the Big Bang. Electrons and positrons annihilate each other when they meet, so when this canceling out was complete some electrons remained. These remaining electrons become the basis of all the matter now in the Universe. Were the initial numbers of electrons and positrons equal, we would have no galaxies, no stars, no planets, nor even any gases.

If the early Universe had expanded at a slightly different rate it would have ended up very differently: a fraction slower and it would have rapidly collapsed in upon itself to form a black hole; a fraction faster and galaxies would never have had the chance to condense.

If the fine-structure constant of nuclear physics were different by a very small amount, the rate at which hydrogen converts into helium would be significantly changed. If the rate were a little slower, the Universe would have remained predominantly hydrogen; slightly faster and it would have become predominantly helium. Either way, stars as we know them would not have evolved.

If the ratio of the masses of the electron and proton were as little as one percent different, it would have been impossible for complex molecules to form.

If the nuclear forces that bind atomic particles together had varied in strength by more than two per cent, no heavy elements would have been formed. There would have been no basis for life.

And, if the gravitational force had been a fraction larger there would have been no convection within stars; no thermal instabilities leading to supernova explosions; no heavy elements scattered into space; and no evolution of matter towards more and more complex forms.

Vary just one of these factors and the Universe as we know it would not exist. Is it all a tremendous series of flukes? Or is the whole Universe, like Gaia, somehow set up so that living systems can evolve? If so, could the Universe as a whole be headed towards becoming a single Universal being?

If, over thousands of millions of years, the ten billion galaxies in the Universe not only evolved into galactic super-organisms, but also began to interact and communicate with each other, there might come the final stage of evolution – a Universal super-organism. A seventh level of evolution might then emerge, a level we could call *Brahman*, after the Indian word for the wholeness of the Universe in both its manifest and unmanifest forms.

If this were indeed the final evolutionary development, it would in some respects bring the whole process full circle. Beginning from a unity of pure energy, the Universe would have evolved through matter, life, consciousness, Gaias and galaxies, to a final reunion in Brahman. From a unity of total non-differentiation it would have evolved, through the most multifarious diversities, to a unity of total integration. From Brahman to Brahman.

What then?

The Universe could possibly collapse in upon itself in some kind of 'Big Wumph'. Would that be the end? Or would it just be the end of one cycle of the Universe?

Maybe another Big Bang and another long chain of evolution would follow? Perhaps in the next Universe there would be a very slight change in the physical constants so that the Universe evolved a little differently. Each cycle might be a fresh experiment, a slight improvement on the previous one: an evolving of evolution itself. If so, Brahman would, so to speak, be reincarnated in each fresh cycle, each time becoming a more perfect Universal being. And the ultimate goal of Universe upon Universe might be the enlightenment of Brahman – the perfect cosmos.

In that final union could come the time of which Olaf Stapledon dreamed in his book *Star Maker*:

This final creature . . . embraced within its own organic texture the essences of all its predecessors; and far more besides. It was like the last movement of a symphony, which may embrace, by the significance of its themes, the essence of the earlier movements; and far more besides . . .

And the Star Maker, that dark power and lucid intelligence, found in the concrete loveliness of his creature the fulfilment of desire. And in the mutual joy of the Star Maker and the ultimate cosmos was conceived, most strangely, the absolute spirit itself, in which all times are present and all being is comprised.

Further Reading

The following list includes books mentioned in the text, books that complement the themes developed here and books that have inspired me. Each of them is highly recommended for further reading. So take your pick, and enjoy.

SRI AUROBINDO, *The Life Divine* (Sri Aurobindo Ashram, Pondicherry, India, 1970). Sri Aurobindo's 1,000 page magnum opus, in which he carefully and logically lays out his basic philosophy on the evolution of man and his future ascent into Supermind. It is, however, slow and exacting reading. A good and readable introduction to Aurobindo's ideas are the selection of writings contained in *The Mind of Light*, ed. Robert A. McDermott, (Dutton, New York, 1971).

ITZHAK BENTOV, *Stalking the Wild Pendulum* (Dutton, New York, 1977, and Wildwood House, London, 1978). A creative and holistic view of human consciousness and the Universe, drawing on holography, quantum physics and Transcendental Meditation. Fun and mind-opening.

KENNETH BOULDING, *The Meaning of the Twentieth Century* (Harper and Row, New York 1965.) An economist's visionary analysis of 'the great transition' – the transition to a post-industrial society – and its evolutionary significance.

MARK BROWN, *Set Thinking* (in preparation). The best introduction to mind sets, how they affect us and how to use them constructively. Entertaining and thought provoking.

FRITJOF CAPRA, *The Turning Point* (Simon and Schuster, New York, and Wildwood House, London, 1982.) A thorough and grounded examination of how the present crises in economic, social, political, medical, educational and other spheres are reflections of an outdated world-view. Written concurrently with my own book, Capra's book looks more closely at many of the issues I have raised. Highly recommended.

ADAM CURLE, *Mystics and Militants* (Tavistock Publication, London, 1972). A balanced and thorough look at the old debate between militant action and inner growth, with a penetrating analysis of the role of belongingness-identity.

MARILYN FERGUSON, *The Aquarian Conspiracy: Personal and Social Transformation in the 1980s* (J.P. Tarcher, Los Angeles, 1980, and Routledge & Kegan Paul, and Granada, London, 1981). A broad look at the many ways in which people are currently working towards the 'New Age', and the leaderless but powerful network which is evolving.

JOHN GRIBBIN, *Genesis* (Dent, London, 1981, Oxford University Press, London, 1982). If you want a detailed and fascinating step-by-step account of the history of the Universe from the Big Bang through to today, this is the book for you.

WILLIS HARMAN, *An Incomplete Guide to the Future* (Stanford Alumni Association, California, and San Francisco Book Co., 1976). An excellent analysis of the basic crises facing humanity, arguing for an evolutionary transformation of society as the only viable means of resolving them in the long-term. Well worth reading.

ERIC JANTSCH, *The Self-Organising Universe* (Pergamon, Oxford and New York, 1980). An account from the perspective of dissipative systems of the whole evolutionary process from Big Bang to Gaia, showing evolution as a natural consequence of physical laws. Also recommended as one of the most readable introductions to the theory of dissipative systems.

THOMAS KUHN, *The Structure of Scientific Revolutions* (second edition, Chicago University Press, 1970). The original and definitive work on scientific paradigms and paradigm shifts.

BARBARA MARX HUBBARD, *The Hunger of Eve* (Stackpole Books, Harrisburg, Penn., 1976). An autobiographical account of one woman's attempts to promote evolutionary thinking. An excellent introduction to an influential and seminal futurist.

ALDOUS HUXLEY, *The Perennial Philosophy* (Fontana, London, 1958, and). A now classic collection of passages from mystics, prophets and saints who have approached direct inner knowledge of the Divine, bringing out the common themes which run across cultures and time.

GEORGE LEONARD, *The Transformation: A Guide to the Inevitable Changes in Humankind* (J.P. Tarcher, Los Angeles, 1981). A very readable overview of the changes in consciousness and society from a human potential perspective. Also *The Silent Pulse* (Dutton, New York, 1978, and Wildwood House, London, 1980). Our identity and our interconnectedness; how getting in touch with the underlying pulse can help personal transformation. A book shining with experience.

JAMES LOVELOCK, *Gaia: A New Look at Life on Earth* (Oxford University Press, London and New York, 1979). One of the original creators of the 'Gaia Hypothesis' details the physical, chemical and biological evidence for the suggestion that the Earth is an organism in its own right.

JAMES GRIER MILLER, *Living Systems* (McGraw-Hill, New York, 1978). Miller's 1000 page magnum opus on the general theory of living systems, detailing how the nineteen critical subsystems of life can be found at all levels, from the single cell through to supranational system.

GUY MURCHIE, *The Seven Mysteries of Life* (Rider/Hutchinson, London, 1979, and Houghton Mifflin, 1978). Seventeen years in the writing, and worth every minute of it. A most imaginative and inspired look at life – from crystals to the whole planet – touching on just about everything, and very simply explained. Highly recommended.

FRED POLAK, *The Image of the Future*, trans. E.Boulding (Jossey-Bass, San Francisco, 1973). One of the principal books on the role of social images in shaping the future.

JAMES ROBERTSON, *The Sane Alternative* (Robertson, Ironbridge, England, 1978, and River Basin Publishing Co., St.Paul, Minnesota, 1979). Another book on the crises facing society and the need for a sane, humane, ecological future. A book of action as well as theory, with valuable suggestions for group discussion.

MARK SATIN, *New Age Politics* (Delta Books, New York, 1979). A most thorough look at the many facets of 'New Age' thinking, values and ethics, and the new politics which are emerging. Invaluable.

RUPERT SHELDRAKE, *A New Science of Life: The Hypothesis of Formative Causation* (Blond and Briggs, London and State Mutual Books, New York, 1981.) A challenging book, suggesting that biological systems are regulated by invisible organising blueprints (morphogenic fields). An ingenious explanation of the transference of behaviours across large distances, which if supported will challenge the current biological paradigm.

CARL AND STEPHANIE SIMONTON, *Getting Well Again* (J.P.Tarcher, Los Angeles and Bantam, New York, 1978.) The role which visualisation and meditation can play in the healing of cancer – though the principles apply to any illness.

WALTER STACE, *Mysticism and Philosophy* (MacMillan, London, 1960, and) A thorough and lucid analysis of the writings of mystics and religious teachers bringing out their common core.

OLAF STAPLEDON, *Starmaker* (Penguin, London and New York, 1972). Arguably the most far-reaching science fiction book ever; yet more of an evolutionary projection than science fiction; from humanity to Gaia, to Galactica and beyond. A must. First published in 1937 it is only now becoming widely acclaimed.

RUSSELL TARG AND HAROLD PUTHOFF, *Mind-Reach* (Paladin, London, and Dell, New York, 1978). Some of the most convincing of the recent evidence that we all have latent ESP.

PIERRE TEILHARD DE CHARDIN, *The Phenomenon of Man* (Collins, London and Harper and Row, New York, 1965) is probably his best known work, although it only presents a partial view of his ideas, and some people find it difficult reading. The best general introduction to his thought is *Let Me Explain*, ed. Jean-Pierre Demoulin (Collins, London, and Harper & Row, New York, 1970), which as well as being an excellent primer will also direct the interested reader through Teilhard's many writings.

WILLIAM IRWIN THOMPSON, *Passages About Earth* (Harper and Row, New York, 1973) and *Darkness and Scattered Light* (Anchor/Doubleday, New York, 1978). Two books by the visionary founder of the Lindisfarne Association exploring the possibility of a planetary renaissance.

ALAN VAUGHAN, *Incredible Coincidence* (Lippincott, New York, 1979). The first major collection of synchronicity case histories.

LYALL WATSON, *Lifetide* (Hodder and Stoughton, London, 1979). The growing awareness that 'we are all one' examined from a biological and evolutionary perspective, complete with the author's usual amazing collection of supporting data.

ALAN WATTS, *The Book: On the Taboo Against Knowing Who You Are* (Random House, New York, 1972, and Abacus, London, 1973). Watt's very readable book on the 'skin-encapsulated ego'.

KEN WILBER, *The Atman Project* (The Theosophical Publishing house, Wheaton, Ill., USA, 1980.) 'The theme of this book is basically simple: development is evolution; evolution is transcendence, and transcendence has its final goal Atman, or ultimate Unity Consciousness . . . ' So begins Wilber's book. It's not light reading, but it is probably the most comprehensive investigation of inner evolution in print. A very important book.

Index

225

ARK PAPERBACKS

PURITY AND DANGER

This remarkable book, which is written in a very graceful, lucid and polemical style, is a symbolic interpretation of the rules of purity and pollution. Mary Douglas shows that to examine what is considered as unclean in any culture is to take a looking-glass approach to the ordered patterning which that culture strives to establish. Such an approach affords a universal understanding of the rules of purity which applies equally to secular and religious life and equally to primitive and modern societies.

MARY DOUGLAS

Mary Douglas is a distinguished international anthropologist who is currently Professor of Anthropology at Northwestern University, Illinois.

ISBN 0-7448-0011-0   208 pp, 198mm×129mm

United Kingdom £2.95 net

USA $6.95

Australia $6.95 (recommended)

Canada $7.50

ARK PAPERBACKS is an imprint of Routledge & Kegan Paul and can be ordered through your usual bookshop.

ARK PAPERBACKS

A SHORT HISTORY OF MODERN PHILOSOPHY

This rich survey is a very successful attempt to make the history of modern philosophy cogent and intelligible to the common reader. Dr Scruton sets out to clarify the principal metaphysical, ethical and political attitudes of post-medieval philosophy, while maintaining a contemporary point of view on each of the major figures discussed. The issues discussed by Descartes and his successors are not closed and modern thinking can only benefit from acquaintance with their methods and arguments. No one contemporary school predominates in this history, all are treated fairly.

ROGER SCRUTON

Roger Scruton is a writer with interests in many fields, including music, literature, the fine arts, architecture and politics. He was educated at Jesus College, Cambridge and is a Reader in Philosophy at Birkbeck College, University of London. He is the author of *The Meaning of Conservatism* (Penguin 1980).

ISBN 0-7448-0010-2   304 pp, 198mm×129mm

United Kingdom £3.50 net

Australia $8.95 (recommended)

Canada $9.50

ARK PAPERBACKS is an imprint of Routledge & Kegan Paul and can be ordered through your usual bookshop.

ARK PAPERBACKS

THE FEAR OF FREEDOM

Does modern man really want freedom — or are we intrinsically afraid of it? In this brilliant account, Erich Fromm asks the fundamental question — is the fear of freedom the root of the twentieth century's predilection for one or other kind of totalitarianism? The rise of democracy, while setting men free, also created a society where man feels isolated from his fellows, where relationships are impersonal and where insecurity replaces a sense of belonging. This sense of isolation drives man to a devotion and submission to all-powerful organization from the state.

The work, which is both psychoanalytical and historical, is a fundamental interpretation of our age and its problems.

ERICH FROMM

Erich Fromm studied at the universities of Heidelberg, Frankfurt and Munich and was trained in psychoanalysis at the Psychoanalytical Institute in Berlin. All his life he devoted himself to both practical psychological consultancy and theoretical investigation. His many published works have been both influential and illuminating. He died in 1980.

ISBN 0-7448-0014-5   272 pp, 198mm×129mm

United Kingdom £3.95 net

Australia $9.50 (recommended)

ARK PAPERBACKS is an imprint of Routledge & Kegan Paul and can be ordered through your usual bookshop.

ARK PAPERBACKS

DICTIONARY OF MODERN CULTURE

With over 300 entries from more than 200 contributors, this is the most
comprehensive and informative survey of twentieth century ideas ever
published. You will refer to this dictionary time and time again, not only
for information and facts, but for stimulation and enjoyment. From
Freud to R D Laing, from Proust to Garcia Marquez, from Picasso to
Warhol, from Chaplin to Godard, from Debussy to Stockhausen, from
Shaw to Pinter, from Wittgenstein to Popper, from Durkheim to
McLuhan, from Yeats to Ginsberg, from Wells to Castaneda.

JUSTIN WINTLE

Justin Wintle was educated at Stowe and Magdalen College Oxford,
where he graduated in Modern History. He has worked as a freelance
writer and editor in London, New York and the Far East. His books
include *The Dictionary of Biographical Quotation*, *Makers of
Nineteenth Century Culture* and *The Dragon's Almanac*.

ISBN 0-7448-0007-2   480 pp, 198mm×129mm

United Kingdom £4.95 net

USA $9.95

Australia $11.95 (recommended)

Canada $12.95

ARK PAPERBACKS is an imprint of Routledge & Kegan Paul and can be ordered through your usual bookshop.

ARK PAPERBACKS

THE SPIRIT IN MAN, ART AND LITERATURE

There are different ways of looking at the achievements of outstanding
personalities. In reading this book, the reader will be in touch with some
of Jung's best insights into artistic and literary creation. The essays are
on Paracelsus, Freud, Richard Wilhelm, Picasso and Joyce's *Ulysses*.
There are also two chapters on poetry and literature.

C G JUNG

C G Jung's collected works are published in full by Routledge and Kegan
Paul.

ISBN 0-7448-0008-0   176 pp, 198mm×129mm

Proof only

United Kingdom £2.95 net

Australia $6.95 (recommended)

Canada $7.50

ARK PAPERBACKS is an imprint of Routledge & Kegan Paul and can be ordered through your usual bookshop.

ARK PAPERBACKS

THE PSYCHOLOGY OF THE TRANSFERENCE

The Psychology of the Transference is an authoritative account of that central issue in all analysis, the handling of the transference between analyst and patient. This is made doubly fascinating because Jung does this by drawing on his conceptions of archetypes and man's inner life. The bond between analyst and patient is seen as being analogous to the kinship libido between the alchemist-adept and his 'mystic sister'. The book is one of the finest of Jung's later writings. It contains practical applications to familiar psychological situations in both the clinical context and everyday life.

C. G. JUNG

C. G. Jung's collected works are published in full by Routledge & Kegan Paul.

ISBN 0-7448-0006-4   224pp, 198mm × 129mm.

United Kingdom £2.50

Australia $5.95 (recommended)

ARK PAPERBACKS is an imprint of Routledge & Kegan Paul and can be ordered through your usual bookshop.

ARK PAPERBACKS

WOMEN OF IDEAS
AND WHAT MEN HAVE DONE TO THEM

Women of Ideas and What Men Have Done To Them is a history of
women's thought. This history is the lost intellectual heritage of women
now hopefully rediscovered and set out for all to read. It is not a mere
Who's Who of biographies, but a fascinating study of thoughts and ideas,
suppression and resurgence spanning three centuries. Dale Spender, as
always a provocative writer, has dug into the hidden past and come up
with intriguing individual examples of feminine creativity and intellectual
prowess which were suppressed and swallowed up by a male-dominated
world. Men have removed women from the literary and historical records
and deprived women of the knowledge of their heritage. This is a
reference work of great value and also a stimulating read.

DALE SPENDER

Dale Spender is an Australian Feminist living and working in London.
She is the author of a number of well known books on Feminism and the
editor of the journal *Women's Studies International Forum*. She lectures
and broadcasts all over the world.

ISBN 0-7448-0003-X   About 900pp, 198mm × 129mm.

United Kingdom £4.95 net

USA $9.95

Australia $11.95 (recommended)

Canada $12.95

ARK PAPERBACKS is an imprint of Routledge & Kegan Paul and can be ordered through your usual
bookshop.